**W9-BCG-392**

UNDILUTED HOCUS-POCUS

# UNDILUTED
# HOCUS-POCUS

---

# THE
# AUTOBIOGRAPHY
# OF
# MARTIN GARDNER

---

PRINCETON UNIVERSITY PRESS   PRINCETON AND OXFORD

Requests for permission to reproduce
material from this work should be sent to
Permissions, Princeton University Press

Published by Princeton University Press,
41 William Street, Princeton, New Jersey 08540

In the United Kingdom: Princeton University Press,
6 Oxford Street, Woodstock, Oxfordshire OX20 1TW

press.princeton.edu

Chapter 8 epigraph from "Chicago," by Carl
Sandburg, © Houghton Mifflin Harcourt.

Chapter 21 epigraph © Woody Allen.

Library of Congress Cataloging-in-Publication Data

Gardner, Martin, 1914-2010.
Undiluted hocus-pocus : the autobiography of Martin Gardner.
pages cm
Includes index.
ISBN 978-0-691-15991-1 (hardcover : acid-free paper) 1. Gardner,
Martin, 1914-2010. 2. Science writers–United States–Biography.
3. Mathematical recreations–United States–History–20th century.
4. Mathematics–Social aspects–United States–History–20th century.
5. Science–Social aspects–United States–History–20th century.
6. Journalists–United States–Biography. 7. Magicians–
United States–Biography. I. Title.
QA29.G268A3 2013
793.74092–dc23
[B]
2013016324

British Library Cataloging-in-Publication Data is available

This book has been composed in Baskerville
10 Pro and League Gothic

Printed on acid-free paper. ∞

Printed in the United States of America

1 3 5 7 9 10 8 6 4 2

*For Jim and Amy,*
ONE MORE TIME

We glibly talk of nature's laws
but do things have a natural cause?
Black earth turned into yellow crocus
is undiluted hocus-pocus.

—Piet Hein

# CONTENTS

# CONTENTS

# FOREWORD:
## MAGIC, MATHEMATICS, AND MYSTERIANS

LIKE THE HERO OF HIS CELEBRATED SHORT STORY, THE "no-sided professor," Martin Gardner was an insider and an outsider at the same time. I first met him in the late 1950s at New York's 42nd Street Cafeteria. A magicians' hangout on Saturday afternoons when the magic shops closed, it was a place where kids, serious amateurs, and professional magicians would traipse downstairs for coffee and "what's new." There was always something new: a sheet of rubber you could pass a coin through, a down-on-his-luck gambler who had spotted the boys (there were hardly any girls) handling cards and stepped inside for a handout in exchange for showing them something they hadn't seen before. A typical Saturday had about fifty people scattered around five or more big circular tables. The youngsters (I was thirteen then) sat together. There was a fifteen-to-twenty-five age group and the big guys' table. There were greats like Dai Vernon, Francis Carlyle, and Harry Lorayne, and various visiting pros and savants held court.

Martin was welcome at the big guys' table. He was only ever a serious amateur magician, but he had invented some wonderful original tricks. Perhaps his best was "lie speller," in which a spectator spelled out the name of a thought-of card, dealing one card off for each letter (j-a-c-k-o-f-c-l-u-b-s); at the end, the true thought-of card showed up *even if the spectator lied along the way*. Martin cataloged, compiled, and described "The Encyclopedia of Impromptu Magic," which ran monthly in the beloved *Hugard's Magical Monthly*. With hundreds of additions, a hardbound edition appeared a few years before his death.

Martin could perform a few obscure, difficult tricks really well. One was the invisible cigarette: a match is lit and the performer pretends to light an invisible cigarette; he mimes taking a puff and blows out a large cloud of real smoke. Another was a vanishing knot: a cleanly tied single knot in a handkerchief fades to nothing when pulled tight. He did a classical trick with two paper matches. Held interlinked at the fingertips, the matches just melted through one another. Martin had a high, quiet whisper of a voice. Frank Garcia, a top sleight-of-hand guy back then, would parody Martin whispering, "Come closer; here's a trick with three human hairs." The diligent reader can find descriptions in *Martin Gardner Presents*.

By this time, Martin had started his epic Mathematical Games column in *Scientific American*. Often, tricks, puzzles, and games picked up at the cafeteria made their way into Martin's columns. You might think that the magician's code of secrecy would have him banned, but the brethren loved having their discoveries read by an audience of close to a million readers. He was a center of attention both as a scholar and as a judge of what was new and worthwhile.

Martin was a quiet, nice guy, polite and enthusiastic, even to thirteen-year-olds. He was interested in everything: tricks, jokes (both black and blue), poems, psychology, and the philosophy of magic. Tricks with a scientific or mathematical background really tickled him, like the following. Find a single playing card. Make slight downward bends in two opposite corners. Hold it lightly from above by one hand, the thumb at one corner, pinky at the opposite. With no apparent effort, the card will slowly rotate until it is faceup. There is no "trick" to this, just a light touch. I was shown it in Toronto by Tom Ransom (himself an amazing magician and great puzzle collector). I showed it to Martin in New York the following week. He was fascinated, and I have five letters from him proposing theories of why it works, the physics of the thing. This is one of tens of thousands of things that Martin thought about.

About those letters: I got about twenty letters a year from Martin over a fifty-year period (you do the math). Sometimes these were elaborate descriptions of tricks of his or others. Sometimes they were brief notes or photocopies of letters others had sent him. Once he sent me a long, wonderful letter about how to write a readable article; it was his way of providing constructive criticism of my paper "Statistical Problems in ESP Research" (*Science*, 1978).

Martin interacted. Once, a book about psychic photography, *The World of Ted Serios* by Jule Eisenbud, M.D., had been sent to *Scientific American* for review. Martin got in touch with photographer/magician Charles Reynolds, technical expert David Eisendrath, and me, and arranged for us to go see Serios to try to figure out how he did it. We caught Serios cheating. The resulting review was too hot for *Scientific Ameri-*

*can* to handle, and the story was published in *Popular Photography* (October 1967). This was the first of many joint interactions with Martin in the world of psychic debunking. I got to visit two Stanford Research Institute (SRI) scientists, call them Puton and Offtarget, who were investigating psychic Uri Geller. My detailed reports on sloppy statistics and uncontrolled experiments were woven into Martin's writings; see his two books in the *Confessions of a Psychic* series under the pseudonym Uriah Fuller (they even appeared in a legal settlement that Martin won from Offtarget). Martin often asked for help in checking the validity of statistics and in doping out methods for supposedly amazing feats. His cabinet of experts included magicians James Randi and Jerry Andrus, and the psychologist Ray Hyman. It was a lifelong battle for him. In addition to science and stealth, he introduced an important new tool to the field: the horse laugh. He just wouldn't go along with the academic tradition of treating crank science seriously.

Of course, Martin was a pioneer at researching pseudo-, fringe, and controversial science with his *Fads and Fallacies in the Name of Science*. This treats Flat (and Hollow) Earthers, flying saucers, dowsing, food fads, "throw away your glasses" claims, Dianetics, general semantics, ESP, and much more. The essays are well researched and interesting. The book was a sensation when it was published in 1952; Martin told me he had received more positive feedback on it than on any of his other writings. However, there was a common undercurrent. After compliments on his critiques, he would often get, "You know, there is one thing I don't understand. Why did you include x in your list of crackpots. I think that's pretty serious stuff. . . ." There was no pattern to the choice of x's: each of his twenty-five chapters was hit equally often. It is natural to wonder what topics Martin would take on today, but we don't have to wonder. He kept on tackling pseudoscience through his column in *Scientific American*, through a long series of articles in the magazine *Skeptical Inquirer*, and in a number of books. Take a look at his *Science: Good, Bad, and Bogus* (Prometheus Books, 1990).

On scholarship, Martin was bound by journalistic, not academic, standards. But he was damn careful. Dozens of his letters to me are requests to check a quote in an obscure book (I collect old magic books). One of many stories stands out. I was visiting Martin after he had moved to Hendersonville, North Carolina, and snooping in his library. I came across a large collection of what I will call books of awful popular poetry. There were more than a hundred volumes with titles like *One Thousand Poems for Children*, *Gleams of Sunshine*, *Favorite Heart Throbs of Famous People*, *Tony's Scrapbook* (eleven years' worth), *The World's Best Poetry* (in ten volumes), and so on. I couldn't figure it out, so I asked him: "Martin, why do you have all these books of awful poetry?" He explained, "I did a collection of best-remembered poems for Dover, and I wanted to know what they were." Sure enough, when I looked, Martin's poetry books had been checkmarked and annotated (he wrote ruthlessly in his books). He had made lists and come to reasoned conclusions. The collection, *Best Remembered Poems* (Dover Publications, 1992) contains standards like Blake's "The Tyger," Burgess's "The Purple Cow," Carroll's "Jabberwocky,"

Whitman's "O Captain, My Captain," and so on, but also many wonderful, less-familiar poems such as Browning's "Meeting at Night," Gray's "Elegy," Hood's "The Song of the Shirt," and Wilcox's "The Winds of Fate." Each poet rates an essay of a few pages with little-known background. I found these fascinating. For example, for "Mary's Lamb," Martin relates the following history. The poem, by Sarah Josepha Hale, first appeared in a children's magazine in 1830. It was widely reprinted, and Thomas Edison recited it in 1877 on the first record ever produced. It became controversial when one Mary Elizabeth Sawyer claimed that she was the original Mary, and that the story had actually happened. Henry Ford bought Mary's story, located the schoolhouse, and restored it as a tourist attraction. The schoolhouse is still open to the public.

The reader should be warned: many poems are accompanied by parodies ("Mary had a little lamb, with green peas on the side, and when her escort saw the check, the poor boob nearly died"). Many of these are by Martin's alter ego, Armand T. Ringer. His comments have bite. In a book review, Martin writes:

> At the front of the book, Alexander quotes these enigmatic lines by Keats: "Beauty is truth, truth beauty; that is all we know on earth and all we need to know." Alas, the lines are almost meaningless. They are not all we know or need to know. Moreover, there are true mathematical theorems that are ugly, and there are beautiful "proofs" that are false. T. S. Eliot surely spoke for most literary critics when he called Keats' lines "A serious blemish on a beautiful poem."

My title is a twist on Martin's *Mathematics, Magic, and Mystery* (Dover Publications, 1956), and it is time to get to mathematics. I wrote the following blurb for one of his books:

> Warning: Martin Gardner has turned dozens of innocent youngsters into math professors and thousands of math professors into innocent youngsters.

I was one of those youngsters. Meeting Martin at thirteen gave me entry into his columns and a peek at mathematics itself. Some mathematicians find the phrase "mathematical recreations" silly. They think there is very little real mathematics in the tricks and puzzles Martin championed. I think that they are being silly. There are mathematical questions everywhere. Most of these are too difficult for the rigid tools that constrain academics, who tend to probe deeper and deeper because "that's where the light is shining." Taking a game, puzzle, or trick and shining your own creative light on it is hard work. There are geniuses in this direction. Martin's constant correspondent John Conway is singularly talented along these lines. Dozens of Conway's games and puzzles saw their first appearances in Martin's column. Perhaps most famous is Conway's "Game of Life," a simple set of rules for allowing pieces on a two-dimensional grid that has a universal Turing machine built into it. This started the modern cellular automaton craze, which has kept logicians, computer scientists, and housewives fascinated for fifty years. Conway created a fantastic world of numbers and games, assigning games to numbers and numbers to games. By numbers, I mean not just 1, 2, 3, . . . , but negative, complex, ordinal, cardinal, infinitesimal, and in some sense "all numbers great and small." The associated games are elaborate extensions of "take-away games" such as Nim: there is a pile of matches on the table; you take some according to a rule, then I do, and so on, and the last to take any wins. These "games" have the deepest mathematical content; probably they will really come to life in the next century.

I think I was present the first time that Conway visited Martin at Hastings-on-Hudson. (Martin lived at 10 Euclid Avenue and liked to point to the parallel coincidence that the great probabilist William Feller lived on Random Road in Princeton.) I knew how to get from Manhattan to Hastings, and Martin asked me to come up on the train, showing Conway the way. I was about sixteen and didn't know any mathematics. Conway needed to have an audience and asked me to play "the unhearing ear." He thought he had seen through the thicket of a difficult proof and asked me to listen as he went through it out loud. "Just nod and pretend you're following, that will be a big help." His arguments were clear, and some parts I thought I could actually follow. I occasionally asked a question, but after a while this annoyed him and I was told to just nod. The rest of the afternoon was similar: Martin and Conway had a rich set of common knowledge that was opaque to me. But Martin and I had a common background in esoteric magic, and we all enjoyed ourselves. All of this opened mathematics to me as a living subject.

Martin helped me in more direct ways. He encouraged me to go back to school after my ten years on the road as a magician. He made suggestions for background when I was stuck on homework, wrote letters of recommendation for graduate school, published some early mathematical work as a problem in his column, and kept me constantly informed of math developments he thought were interesting. As time went on, I helped him debug statistical issues, and he listened patiently when my mathematical development went professional. He became my unhearing ear.

Mathematicians weren't the only ones who fought recreational math. Martin wrote often about incorporating it into the K–12 curriculum. Educators have not been receptive, either. There are arguments on both sides, but the community hasn't seriously tried.

One person who has tried is the Stanford statistician Susan Holmes. She was teaching master's students in psychology who had failed at math and had to pass comprehensive exams. She used Martin's *Aha! Insight* and *Aha! Gotcha* (W. H. Freeman & Co., 1978 and 1982) as texts. These are filled with seemingly simple trick problems and paradoxes. She got the students really working by having them study *their* psychology when blocked by a problem. I just looked over these books. *Gotcha* begins with the liar paradox, "This statement is false," running it through real life, politics, literature, and logic. All of Martin's chapters strike me that way on repeated rereading. For me, it is his greatest magic trick.

On to mysterians. Martin's academic training was an undergraduate degree in philosophy from the University of Chicago. He almost majored in religion and had a lifetime interest in both. His novel *The Flight of Peter Fromm* (William Kaufmann, 1973) is a thinly disguised dissection of avant-garde religion as developed in his time at Chicago. He told me a story about his Chicago days. After the war, around 1948, the Vienna

Circle philosopher Rudolf Carnap was visiting Chicago and gave some public lectures on his current thoughts. The next day, Martin was browsing in the wonderful seminary bookstore, and one of his favorite philosophy professors spotted him. The professor had seen Martin at Carnap's lecture and asked him what he thought. Martin hadn't followed well, but the professor drew him out, and together they outlined Martin's views. About that time, Carnap walked in. The professor greeted him and introduced Martin. "We were just talking about your lecture last night. Martin, why don't you tell him what you were telling me." Thinking it was approved, Martin launched into the detailed version he had settled on. Carnap was extremely tall, and formally dressed, and Martin was a little guy, dressed as a 1948 undergrad. Carnap listened for a minute or two and angrily disagreed: "I can see that you have no background in philosophy whatsoever. Everything you have said is completely off. . . ." Martin apologized and slunk away.

Martin wasn't put off. He sat through Carnap's course on philosophy of science and years later (1958) contacted Carnap to inquire about writing it up. Carnap had moved on to UCLA, tape-recorded the course, and he and Martin made it into a book, *Philosophical Foundations of Physics* (Basic Books, 1966). Carnap's later work overlaps mine in foundations of probability. He wasn't dumb, but he was deadly dull. Martin's version is full of life, and recommended reading.

Martin put his philosophical ideas together in the wonderful *Whys of a Philosophical Scrivener* (Quill, 1983). The chapters are wonderful, lively overviews of basic questions from "Why I am not a solipsist"

through "Why I cannot take the world for granted."
In each, Martin explains the issues "in English," outlines the philosophical community's views, and makes
sure to tell us what he believes and why. One chapter
of it shocked me. Martin was such a dedicated skeptic
that I was sure he was an atheist. Chapter 13, "Faith:
Why I Am Not an Atheist," set me straight. Of course,
the discussion is subtle and roundabout, but there is
no denying Martin's acceptance of things (a personal
God, human immortality) that I find fanciful. He was
happy to discuss it at length. A wonderful coda: the
book had a scathingly negative review in the *New York
Review of Books* (December 8, 1983). The knowledgeable reviewer, George Groth, tore the book to shreds:
"He defends a point of view so anachronistic, so out
of step with current fashion, that were it not for a
plethora of contemporary quotations and citations,
his book could almost have been written at the time
of Kant, a thinker the author apparently admires." The
detailed hatchet job goes on for many columns. The
last sentence has a surprise: "George Groth is one of
Martin Gardner's pseudonyms." All told, if I had to
form my own opinion on any of his topics, I do not
know a finer place to start.

Martin remained fascinated by philosophy. A few
years before the end of his life, he began to talk about
being a "mysterian." (See his essay "I Am a Mysterian," following the preface of this book.) The way he
put it to me is simple:

Most anyone would accept as a fact that chimpanzees
are incapable of understanding quantum mechanics. I
feel the same way about humans understanding the hard

questions that stump them: how did the world begin, does God exist, and so on. These questions are so far above our natural human prowess that to fret about them seems as ridiculous as insisting that a dog understand general relativity.

It's defeatist in some ways but a great relief in others. I got a minor dose of this recently from some physics colleagues. I sometimes coteach in the physics department. I am a mathematician and always looking for clear argument (proof) from believable hypotheses to a desired conclusion. Physics, or chemistry, or biology, isn't like this. There are standards of argument: if some formal series expansion agrees with computer simulations and data from experiments, that's a pretty good argument. It bothers me to accept things without more formal proof. In our course, we set aside one hour a month for discussions on "the value of proof." I thought I was going to kill on this topic. Not so fast. They got me with, "Listen, you don't know how the radio works, do you?" "Nope." "But you turn it on and it works fine. We can't prove that matter at low temperature is approximately crystalline. But you sprinkle salt crystals on your food. Do you have to be able to prove things in order to use them?"

These questions are not as deep as Martin's, and I won't stop trying, but having his mysterian philosophy as an option is a gift. It's one of many. Martin is gone, but his depth and clarity will illuminate our world for a long time. Pick up anything he wrote. You'll smile and learn something.

*Persi Diaconis*
STANFORD, CALIFORNIA

# PREFACE

THE AMERICAN PHILOSOPHER WILLIAM JAMES ONCE said he couldn't understand how anyone could read the Bible from cover to cover and believe it was the word of God. The Old Testament's Jehovah is an angry, cruel tyrant. He drowns every man, woman, and child, and their pets, except for Noah and his family. He turns Lot's wife into a pillar of salt because she disobeys him by looking back at Sodom and Gomorrah while his angels are demolishing both towns. When Moses's two nephews mixed incense improperly for an animal sacrifice, God so disliked the smoky smell that he killed both boys with lightning bolts.

Jesus's God was more merciful. He only tortured eternally such evil men as the extremely wealthy. Paul's God was worse. He punished forever (forever!) those who held incorrect *beliefs* about Jesus and his Resurrection!

The best-known remark of stand-up comedian Lenny Bruce was that people are leaving their churches and going back to God. What follows here is a rambling autobiography of one such person—me.

*Martin Gardner*
NORMAN, OKLAHOMA
SPRING, 2010

# PROLOGUE: I AM A MYSTERIAN

OUR BRAIN IS A SMALL LUMP OF ORGANIC MOLECULES. It contains some hundred billion neurons, each more complex than a galaxy. They are connected in over a million billion ways. By what incredible hocus-pocus does this tangle of twisted filaments become aware of itself as a living thing, capable of love and hate, of writing novels and symphonies, feeling pleasure and pain, with a will free to do good and evil?

Let me spread my cards on the table. I belong to a small group of thinkers called the "mysterians." It includes Thomas Nagel, Colin McGinn, Jerry Fodor, also Noam Chomsky, Roger Penrose, and a few others.

We share a conviction that no philosopher or scientist living today has the foggiest notion of how consciousness, and its inseparable companion free will, emerge, as they surely do, from a material brain. It is impossible to imagine being aware we exist without having some free will, if only the ability to blink or to decide what to think about next. It is equally impossible to imagine having free will without being at least partly conscious.

In dreams one is dimly conscious but usually without free will. Vivid out-of-body dreams are exceptions. Many decades ago, when I was for a short time taking tranquilizers, I was fully aware in out-of-body dreams that I was dreaming, but could make genuine decisions. In one dream, when I was in a strange house, I wondered if I could produce a loud noise. I picked up a heavy object and flung it against a mirror. The glass shattered with a crash that woke me. In another OOB dream I lifted a burning cigar from an ashtray and held it to my nose to see if I could smell it. I could.

We mysterians are persuaded that no computer of the sort we know how to build—that is, one made with wires and switches—will ever cross a threshold to become aware of what it is doing. No chess program, however advanced, will know it is playing chess any more than a washing machine knows it is washing clothes. Today's most powerful computers differ from an abacus only in their power to obey more complicated algorithms, to twiddle ones and zeroes at incredible speeds.

A few mysterians believe that science, some glorious day, will discover the secret of consciousness. Penrose, for example, thinks the mystery may yield to a deeper understanding of quantum mechanics. I belong to a more radical wing. We believe it is the height of hubris to suppose that evolution has stopped improving brains. Although our DNA is almost identical to a chimpanzee's, there is no way to teach calculus to a chimp, or even make it understand the square root of 2. Surely there are truths as far beyond our grasp as our grasp is beyond that of a cow.

Why is our universe mathematically structured? Why does it, as Stephen Hawking recently put it, bother to exist? Why is there something rather than nothing? There may be advanced life-forms in Andromeda who know the answers. I sure don't. And neither do you.

*Martin Gardner, August 2007*
EXCERPTED FROM "DO LOOPS EXPLAIN CONSCIOUSNESS? REVIEW OF *I AM A STRANGE LOOP*," *NOTICES OF THE AMS* 54, NO. 7 (AUGUST 2007). PUBLISHED BY THE AMERICAN MATHEMATICAL SOCIETY.

UNDILUTED HOCUS-POCUS

# 1

## EARLIEST MEMORIES

I HAVE ALWAYS LOVED COLORS. ALL COLORS. TO ME THE
ability to see colors is one of God's great blessings.
(Yes, Virginia, there is a God. In my final chapter I
explain why I call myself a philosophical theist.)
Searching my brain for the earliest event I can recall,
the best I can come up with is a memory involving
colors while I was being carried in my father's arms on
a fine autumn day in Tulsa. The ground was covered
with dead maple leaves. I pointed to a leaf and somehow indicated I wanted it. My dad picked it up and
handed it to me. It was gorgeously blazing with reds
and browns and yellows.

My mother, too, was fond of colors. When she was a
kindergarten teacher in Lexington, Kentucky, trained
in the Montessori Method, she liked to teach her children the names of colors. I remember when she made
for me six balls of yarn, three bright with the primary
colors, three with the secondary colors. She would
point to objects in a room and ask me to name their
color. Late in life, when she studied art under Adah
Robinson, at Tulsa University, she reveled in the colors of dozens of still lifes she painted.

Miss Robinson was well known in Tulsa as designer
of the Boston Avenue Methodist Church, which we

attended. She also designed the interior of Tulsa's First Church of Christ, Scientist. Her oil portrait of my mother is owned by Tulsa's Gilcrease Museum.

I remember one day, when I was a child in bed with some illness, my mother brought a box of watercolors to the bed and on a sheet of paper painted a picture of a sunset. I can still vividly recall its glowing colors.

Hanging in our large house at 2187 South Owasso Street, Tulsa, were several watercolors by the Kentucky artist Paul Sawyier, a painter my mother greatly admired. Many years later I sold to a gallery in Frankfort, Kentucky, reproduction rights to a Sawyier picture of a covered bridge. He is Kentucky's most famous artist. There is a room devoted to his work in Frankfort's capitol building. You can buy a huge volume about his paintings.

An indication of my mother's love of colors was her enormous delight in seeing a rainbow. She always looked for one if there was a shower accompanied by sunshine, especially from a sun low in the sky. She would rush outside to look for a bow. If there was one, she would hurry to the phone and call a dozen friends, urging them to go outside to see the bow. To paraphrase a familiar couplet by Wordsworth:

> Her heart leaps up when she beholds
> A rainbow in the sky.

Now that I am an old man, my heart still leaps up when I, too, see a rainbow. It made a high leap one morning when I saw a secondary bow. The wonderful thing about a rainbow is that it is not something "out there" in the sky. It exists only on the retinas of eyes or on photographic film. Your image in a mirror is similar. It's not a thing behind the looking glass. By the way, what does a mirror look like when there's no one in the room? And why does a mirror reverse left and right but not up and down?

Over the living room's fireplace, in our Owasso Street home, was an oil painting by a famous Dutch artist, Bernard Pothast (1882–1966). It showed a mother and child blowing bubbles. I remember admiring the shifting colors of bubbles blown by me and my mother, and trying to catch them. One of Mother's favorite songs was "I'm forever blowing bubbles, pretty bubbles in the air. They fly so high, they nearly reach the sky, then like my dreams they fade and die. Fortune's always hiding, I've looked everywhere. I'm forever blowing bubbles. Pretty bubbles in the air." I have also not forgotten the tune, which I enjoy playing on my musical saw.

Yes, I play the saw. One of G. K. Chesterton's familiar aphorisms is that if anything is worth doing, it is worth doing badly. I play the saw badly. Like Sherlock Holmes and his violin, when I have nothing better to do, I take down my saw from a wall hook, along with a felt-tipped wooden mallet, and relax for half an hour tapping out familiar tunes. There are of course hundreds to pick from, including gospel golden oldies with crude lyrics I have been unable to forget.

Bouncing a left leg adds vibrato to the saw's pure tones. I have yet to advance, perhaps never will, to using a cello bow instead of a mallet to keep the saw vibrating. I doubt if many readers know that Marlene Dietrich was a saw virtuoso. She even gave concerts! In my crazy novel *Visitors from Oz*, I have Dorothy playing the saw on an Oprah Winfrey show, having learned how to play it from Kansas farmhands. The Tin Woodman provides bass by thumping his hollow chest with his tin fists.

L. Frank Baum, who created Oz, is one of my literary heroes. He was so fond of colors that he divided Oz into five regions, each with a dominant color. On the east is Munchkin territory where the color is blue. To the west is Winkie country where the dominant color is yellow. Purple tinges the wild northern region of Oz, and red dominates the southern Quadling country where Glinda lives. In the center of Oz is the green Emerald City. (If I ever write another Oz book, I'll introduce Orangeville where the dominant color is orange.) I persuaded my parents to buy all of Baum's fourteen Oz books as well as all his non-Oz fantasies, some of which—*Sky Island*, for example—I consider better written than many of his Oz books.

When I was a boy, I was so fond of green that my mother had my bedroom walls papered green. To this day when I see a small girl dressed in blue, I think of her as a munchkin. If I'm served bright green Jell-O, I can't help momentarily imagining I'm in the Emerald City.

As an adult I had the immense pleasure of joining Jack Snow, author of two Oz books and *Who's Who in Oz*, and Justin Schiller in founding the International Wizard of Oz Club. (See the chapter on "How the Oz Club Started" in my book *The Jinn from Hyperspace*.) It was Justin who at age fourteen started what he called the *Baum Bugle*. It consisted then of several mimeographed sheets stapled together. On its masthead I'm listed as Chairman of the Board of Directors! Today the Oz Club holds four annual conventions in four

American cities and publishes a handsome scholarly quarterly still called the *Baum Bugle.*

At about the time the Oz Club got underway, and hundreds of members of all ages suddenly discovered they were not alone in their love of the Oz books, American critics and librarians had almost no interest in Baum. One woman scholar wrote a book of more than four hundred pages on the history of juvenile literature. It contained not a single mention of Baum! While the Judy Garland movie was introducing millions of children to Oz, the head of Detroit's libraries proudly announced that he considered the Oz books so unsuitable for children that he did not allow such books in any city library! This so infuriated a Detroit newspaper that it serialized *The Wizard of Oz,* heading each episode with a statement that this is the book your child can't get in any Detroit library. After a raft of articles about Baum began to appear in prestigious journals, the librarians began to change their minds. A turning point came when Columbia University's library, headed by Roland Baughman, a Baum enthusiast, sponsored an exhibit of first editions of all of Baum's books. The catalog of that exhibit is now a rare collector's item.

Only a few years before Columbia's exhibit I had published in *Fantasy and Science Fiction,* edited by Oz buff Anthony Boucher, a two-part biography of Baum. It listed for the first time a fairly complete bibliography of all of Baum's books. This included books published under various pseudonyms, such as his series of books for girls written under the name of Edith Van Dyne. Before my biography ran, I visited a used bookstore in New Jersey where I bought all the Van

Dyne books for twenty-five cents each. After my biography appeared, book dealers who had never realized they had Baum books in stock, jumped the prices to twenty dollars.

I learned to read by looking over my mother's shoulder while she read aloud *The Wizard of Oz*. I simply followed the words as she spoke. This created a problem for me in the first grade. A teacher would hold up cards with words such as *dog* and *cat*, and I would be the first to call them out. The teacher forced me to keep quiet while she worked with other children. Of course this meant I had to sit in silent boredom.

As an adult I wrote introductions to six Dover paperbacks of Baum's best fantasies about enchanted lands other than Oz. Finally I wrote an Oz book myself, *Visitors from Oz*. It is not for children, but aimed at adults familiar with the Oz books. It tells of the adventures of Dorothy, Scarecrow, and Tin Woodman in the Gillikin region of Oz, followed by their visit to New York City to publicize a new musical about Oz. I reveal for the first time that Lewis Carroll's Wonderland is actually underground in Oz, that Mary Poppins lives in Oz, and that the exiled Greek gods have found a home on a small Mount Olympus in Oz. The *New York Times* called my book "a poor thing of a novel," but to my surprise (because Baum is little known in England) the London *Times Literary Supplement* gave the novel a long and favorable review.

Another of my literary heroes who was fond of colors is Gilbert Chesterton. One of his best short stories, "The Coloured Lands," first appeared in a posthumous book by the same name. In an introduction for a Dover reprint, I summarized the story this way:

"The Coloured Lands" is a short tale about a strange young man who lets a boy named Tommy look through four spectacles with colored glass that turn everything into blue, red, yellow, or green. The man tells Tommy that when he was a child he had been fascinated by colored glasses, but soon tired of seeing the world in single colors. In a rose-red city, he explains, you can't see the color of a rose because everything is red. At the suggestion of a powerful wizard, the man was told to paint the scenery any way he liked:

> "So I set to work very carefully; first blocking in a great deal of blue, because I thought it would throw up a sort of square of white in the middle; and then I thought a fringe of a sort of dead gold would look well along the top of the white; and I spilt some green at the bottom of it. As for red, I had already found out the secret about red. You have to have a very little of it to make a lot of it. So I just made a row of little blobs of bright red on the white just above the green; and as I went on working at the details, I slowly discovered what I was doing; which is what very few people ever discover in this world. I found I had put back, bit by bit, the whole of that picture over there in front of us. I had made that white cottage with the thatch and that summer sky behind it and that green lawn below; and the row of the red flowers just as you see them now. That is how they come to be there. I thought you might be interested to know it."

Chesterton's fiction glows with color words. There are beautiful descriptions of sunrises and sunsets. He liked to put red hair on the women in his novels, even occasionally on men. Trained at a commercial

7

art school—G. K. never went to college—he loved to draw with colored chalk on brown paper. You'll find some of his color sketches in *The Coloured Lands*. One of his finest essays, "The Glory of Gray," is about how gray backgrounds enhance the brilliance of any color. I have always regretted that G. K. never read an Oz book. I think the colors of Oz would have delighted him as much as they delighted me.

Only a few other memories of my very early years float to mind as worth telling. My parents' first house, on the north side of Tulsa, was a tiny one. I remember nothing about it except that it had an outdoor pump at which my mother drew water. Tulsa then was a small village without running water. My only memory of the house is of a cook killing a chicken in the backyard by snapping off its head, and how the poor bird flapped headlessly about the grass for several minutes.

Our second and larger house was on North Denver and is still there. I have only dim memories of it, such as standing on a top step and touching the ceiling with a hand. I remember falling off a sofa and breaking my left wrist. I recall being in a hospital for circumcision, the reason for which puzzled me at the time. And I can remember having my tonsils removed, and enjoying ice cream for several days.

I can't resist including an amusing incident that occurred before I was old enough to remember it. I know of it only because I later heard my father tell it. My parents and I were visiting my father's brother, Uncle Emmett, in Louisville, Kentucky. He was one of the state's earliest psychiatrists, with a practice so successful that he founded and ran the city's first mental hospital. He was a tall, handsome man with red hair and a curious tongue that was crisscrossed with deep furrows. Like my father he had a great sense of humor. He enjoyed hearing and telling jokes about psychiatrists and their patients. For example, a patient told his psychiatrist he couldn't sleep at night because of the smell of a goat he kept in his bedroom with all windows closed.

"Why don't you open a window?" the psychiatrist asked.

"And let all my pigeons out?"

One day Uncle Emmett started to tell me a riddle about a duck in front of two ducks, a duck behind two ducks, and a duck in the middle of two ducks. How many ducks were there? Emmett forgot to withhold the answer. He began by saying, "There were three ducks"; then he caught his mistake and broke into loud guffaws.

That, however, is not the event my dad liked to tell. One night on our visit to Louisville I shared a bed with my uncle. During the night I awoke with a strong urge to urinate. Uncle Emmett, half asleep, took an empty glass off a bedside table and held it while I relieved myself. He put the glass back on the table and we both went back to sleep. In the morning we found the bed soaked. The glass was upside down!

Emmett, I should add, was far ahead of his time in his low opinion of Freud. I once asked him what he thought about the then popular books by an American psychoanalyst, Karen Horney. He had never heard of her!

# 2

## LEE SCHOOL

You were my queen in calico.
I was your bashful, barefoot beau.

*—1907 song by Will Cobb and Gus Edwards*

I ATTENDED GRADES ONE THROUGH SIX AT LEE SCHOOL, a redbrick building within walking distance of where I lived at 2187 South Owasso. The house was large enough to accommodate my grandmother Lucy; her brother, Uncle Owen; my younger brother, Jim; and later my sister, Judith. Lee School is still there. I have happy memories of my teachers, especially Mrs. Polk, who had all her students reading and memorizing large chunks of popular verse. I can still recite all of Longfellow's "The Day Is Done" and the first stanza of Noyes's "The Highwayman":

The wind was a torrent of darkness
    among the gusty trees.
The moon was a ghostly galleon
    tossed upon cloudy seas.
The road was a ribbon of moonlight
    over the purple moor,
And the highwayman came riding—

Riding—riding—
The highwayman came riding, up to the
old inn-door.*

Mrs. Polk's enthusiasm for poetry was catching. Alas, we also had to memorize one of her own poems, which were close to doggerel. I still recall how it began:

When I see a star with its mellow light aglow
I think of Him who placed it there
A million years ago.

Mrs. Polk was an avid reader of mystery fiction. At the time *The Bishop Murder Case*, by S. S. Van Dine, was being serialized in a periodical, and she and I had been reading the installments. She phoned one day to ask if I had figured out who the serial killer was. I hadn't, but she had, and she proved to be absolutely correct.

One afternoon when Mrs. Polk visited our home, I showed her what magicians call a four-ace trick. The aces are at the bottom of four piles of cards of four cards each. An onlooker selects a pile. The aces then vanish from the other three piles and appear in the chosen pile. The trick has endless variations.

My version used three double-faced cards—aces that become other cards when turned over. Mrs. Polk was much impressed. I mention this here because it is my earliest memory of having shown a magic trick to anyone outside my family, and because it shows how young I was to be interested in conjuring.

* For my parody of this poem, in which Bess, the landlord's black-eyed daughter, shoots her lover instead of herself, see my book *Favorite Poetic Parodies*. It appears under the byline of Armand T. Ringer, an anagram of my name.

A sad classroom incident has stuck in my memory. I think it was at Lee School but it may have been later at Horace Mann, a junior high school I attended. For some reason or other a student had walked to the door. The teacher, new on the job and vainly trying to restore order, gave him a shove. His head banged against the door and shattered its glass upper half. I remember feeling sorry for the bewildered teacher, who had no experience in keeping a class in order, and ashamed that I had participated in the ruckus. Of course the poor woman was fired, replaced by a stern teacher who soon had us all under her thumb.

Down the street from 2187 lived a man with curious ears. He was able to fold them and push them into his ear holes, where they would stay folded until he wiggled his ears; then they would pop out to normal. Living in back of this man's yard was a large black dog named Rajie. Rajie loved to visit our house, which had a large yard and a tennis court surrounded by a wire fence over which tennis balls would occasionally fly to get lost in foliage. Rajie loved to look for lost balls. When he found one, he would bring it to someone, begging the person to toss it. He would then gallop off to catch it, often on a high bounce, and bring it back for endless repeats. I got to know Rajie well enough to consider him a friend. He was the most intelligent dog I ever knew.

One afternoon my father placed a tennis ball under an inverted bucket. Rajie did his best to overturn the bucket by pushing it with his nose, but the bucket refused to turn upright. Dad then picked up one of Rajie's paws, placed it on the bottom of the bucket, and turned the bucket over. When Dad put the ball back under the bucket, Rajie immediately turned it over with a paw.

For some reason that I can't recall, I was once in the backyard where Rajie lived in a large doghouse. I peeked inside. The dog was off somewhere, but I found his house filled with dozens of tennis balls he had collected!

I sat in Lee School classes with two science teachers. One, an attractive young blonde, knew her science well. I remember being surprised when she told the class that the stalactites and stalagmites in Carlsbad Caverns proved that the earth was much older than the ten thousand years of Genesis. In those days public school teachers, even more than now, were afraid to question in any way the Bible's historical accuracy.

The other teacher, a middle-aged woman with no training in science, assured the class we would never land spaceships on the moon. Rockets simply won't work, she said, in the vacuum of space. I had some arguments with her because I had been reading lots of science fiction, including novels by Jules Verne and H. G. Wells, and I took for granted that space travel was inevitable. In high school I became a charter subscriber to Hugo Gernsback's *Amazing Stories*, the world's first magazine devoted entirely to science fiction. Had I saved the first year of copies, they would today be worth a tidy sum, but I gave them all to my high school physics teacher, M. E. Hurst, about whom I will have more to say in the next chapter.

Before subscribing to *Amazing Stories*, I also received monthly a copy of Hugo Gernsback's wonderful magazine *Science and Invention*. It not only published science fiction before *Amazing Stories* did, but its wild speculations about the future of science were a delightful mixture of hits and misses. I remember a great cover showing what a Martian would look like. Another cover, one of the hits, showed a couple kissing with wires connecting parts of their bodies to their brains and to equipment measuring their responses to the kiss. The cover illustrated an article on how science was starting to investigate sex. Other cover hits were pictures of helicopters helping build skyscrapers, and the use of flamethrowers in wars. Covers were devoted to articles debunking pseudosciences such as astrology, spiritualism, and perpetual motion. One fascinating cover depicted a scene in which you were asked to count all the scientific mistakes, such as a rainbow with colors in the wrong order, jets of water following wrong curves, and so on.

There were lots of articles about magic. The magician Joseph Dunninger had a monthly page devoted to magic. I recall a cover showing Dunninger cutting a lady in half with a giant buzz saw. Another cover illustrated an article on how to build a theramin, an electric musical instrument that is controlled purely by waving one's hands near two antennas—never touching them. Walter Gibson, creator of *The Shadow* and author of many books about magic, contributed articles on tricks with coins, handkerchiefs, matches, cards, and so on. Pages were devoted to the latest inventions, and to outlandish patents.

Gernsback published forty short stories by a high school science teacher named Clement Fezandie. They were about Dr. Hackensaw's scientific discoveries. Not one of his stories has since been reprinted even though they pioneered dozens of themes later exploited by SF writers. I once tried to interest publishers in letting me do an anthology of the best from *Science and Invention*, but there were no takers.

For years the covers of *Science and Invention* were printed on gold paper to symbolize the golden age of science. Gernsback's editorials alone were wise and worth reprinting. In the twenties he actually ran a television broadcasting station in New York City! The screen was about the size of a postcard, and you had to build your own receiving set. Its pictures were produced by a rotating disk with spiral holes in synch with a similar disk that scanned the scene to be transmitted. For my tribute to Gernsback see the last chapter of my book *From the Wandering Jew to William F. Buckley, Jr.*

My mother had two relatives in Lexington who were growing old, in ill health, with fading eyesight, and living in poverty. They were Lucy, my mother's mother, and her brother Owen. My father, wealthy from his oil wells, agreed to moving Lucy and Owen to Tulsa to occupy a house big enough to give them private rooms. Years later, when Aunt Annie in Lexington was dying of cancer, my mother let her live in the house's guest room until she passed on.

The large stucco house, which my dad bought from the family who had the house built, still stands at 2187 South Owasso Street on Tulsa's south side. It contained five bathrooms, and a third floor for servants, complete with kitchen and bath. The yard was extensive, with a garage that had an apartment on its second floor, and a cement-floor tennis court surrounded by a high wire fence.

After my sister, Judith, was married, and living in the East, she noticed one day on a visit to Tulsa that the old house was vacant and up for sale. She visited the house and was much amused when the agent took a group to a spot where she said there had once been a swimming pool. It was the spot where the tennis court had stood. Judy did not correct the agent or tell her she had once lived in the house.

Around 2008 I had the eerie pleasure of going through the old house on South Owasso. It came about as follows.

Dana Richards, a computer scientist at George Mason University, for years has kept a detailed bibliography of everything I have had published, even including letters to editors. He is also working on a biography. For the biography he wanted a photograph of me at the front door of the Tulsa house, so he drove me from Norman to Tulsa to get the picture. While he was snapping the camera, the lady of the house came out to see what was going on. After Dana explained, she invited us into her home and for half an hour allowed me to roam through it. Of course this brought back a flood of happy memories.

The house had many previous owners who did extensive redecorating, both inside and out. The large yard had been chopped into pieces that sold individually. The tennis court and garage no longer exist. Inside the house the biggest change was the installation of an elevator that ran from the basement to the third floor.

Originally, the third floor had been designed for servants, complete with bathroom and kitchen. There was a back stairway from the third floor to the main kitchen. It was closed off from the rest of the house so residents of the top floor could come and go without passing through lower rooms. I explained to the lady of the house how my brother and I played a game on the hidden stairway. The game was this. Standing at the top of the stairs, we each held a tennis ball on which our initials were inked. We tossed our balls, and sometimes threw them as hard as we could, down the stairs, and the ball that went the farthest would be the winner. On rare occasions a ball would bounce all the way to roll into the basement. The lady of the house was much amused. She said she would have to teach the game to her two sons.

The house when I lived there had a cockroach problem. By night they would invade the kitchen, then go back to the basement in the morning. Eventually they had to be professionally exterminated. The basement was large, dark, and dank, crammed with trunks, old furniture, spiderwebs, and a hundred other things. I recall one day at dinner when my father suddenly asked my mother, "Whatever happened to my crib?" She explained that it was in the basement.

Gran, as I called my grandmother, shared Mother's Protestant faith, firmly convinced that the Bible was the word of God. Both she and my uncle had only a high school education. I never saw either read a book, although Gran enjoyed motion pictures, especially those starring Lon Chaney. Uncle Owen's main reading was the sports pages of the two Tulsa papers. Uncle was fond of baseball and liked to keep up with the major league games. When I was a small boy, he took me to many ball games where the home team was called the Tulsa Oilers. We always sat in the bleachers near third base. Uncle would explain arcane baseball rules, and of course I rooted for the home team while I enjoyed a box of Cracker Jack and a bottle of soda.

Uncle Owen was a quiet man who seldom spoke. My mother told me he had been an accountant. He actually had memorized the multiplication table through 99! If you gave him two numbers, each of two digits, he could at once tell you their product! As for his religious views I haven't the foggiest notion. He lived quietly in his bedroom, occasionally playing a game of solitaire or listening to the radio. He had two main daily tasks. Each day he would see that all the clocks in the house were working and wound if necessary, and make sure the front and back doors were locked. On Sundays he walked to a spot where out-of-town newspapers were sold, and came back with half a dozen papers that carried Sunday comics. I would invite friends over, and we would spend an hour on the floor reading what we called the "funny papers."

One day I heard Uncle Owen sobbing in his room. When I asked why he was crying, he waved me away, asking me to leave. I never found out why he was weeping. He had a male friend in Terre Haute, Indiana, who had recently died. This may have been the cause of his sobbing.

Mother once told me that in his youth Owen liked to walk on his hands. He taught me three magic tricks with a loop of string. One cleverly released a finger ring from the string. Another released a string after it was put through a shirt buttonhole and its ends looped over the thumbs. The third was a way of pulling the string free of the left hand after it seemed hopelessly tangled around the fingers.

I loved Owen and Gran very much. He once told me his favorite poem was "The Burial of Sir John Moore," by Charles Wolfe. I recall him once reciting a short poem that began, "O Burr, O Burr, what have you done? You've slain the noble Hamilton." I regret that I never questioned Gran or Uncle about their life in Lexington.

Jean Craver, a Lee School classmate who lived a few houses from 2187, was my first girlfriend. We would walk home from school together, me carrying her books. Jean frequently visited our yard to join other neighborhood children in hide-and-go-seek games, notably one we called Kick the Can.

Jean's father was the last of a curious breed of men who operated what oilmen called "doodlebugs." These were devices they had invented for indicating whether oil was under a piece of land. Some doodlebug operators actually believed in their contraption, but most were outright crooks. They would visit credulous farmers, then for a modest fee tell them if their property was over oil. I have no idea whether Mr. Craver was a true believer or a swindler. At any rate he earned a comfortable living by sticking wires of his doodlebug in the soil and reading the figures it displayed.

Many decades later I looked up Jean while on a brief visit to Tulsa. Her husband had died in a car accident. We reminisced about our childhoods. I asked if she remembered an elaborate valentine I had sent to her. She remembered it well. I recited her old phone number. To my amazement she recited mine. On a later Tulsa visit Jean invited me to her home to meet her second husband. He impressed me as a handsome good fellow, and I was happy for them both. On our previous get-together she said something nice that I never forgot. She said, "Where have you been all my life?"

I have fond memories of Jean, although we never kissed or even held hands.

# 3

## TULSA CENTRAL HIGH, I

HIGH SCHOOL WAS LIKE FOUR YEARS IN PRISON. I HATED it. With the exception of classes in mathematics and physics, I firmly believe that my years in high school were totally wasted. I particularly disliked history. It seemed concerned only with idiotic kings and queens, and meaningless religious wars that were like the one in *Gulliver's Travels* fought over the proper way to crack an egg. The really important history, it seemed to me, was the history of science. Of all the vast changes in human life, most are the result of the steady progress of science and technology.

Newton did more to alter the world than any king or queen or great military leader. Einstein, sitting alone and thinking, changed the world more than any politician. Remember $E = mc^2$? It explains what happens when an atomic bomb explodes. If Aristotle were alive today, he would have no difficulty understanding our art, music, poetry, literature, even our religions and philosophies. But he would be overwhelmed by skyscrapers, cars, airplanes, most of all by pocket calculators and television.

I remember a high school class in English literature. We had to study some plays by Shakespeare. Reading the bard at seventeen turned me off Shakespeare for years. It was not until I picked up a paperback copy of *Midsummer Night's Dream*, in a sleepy little town in Texas, that I discovered to my delight that Shakespeare was a great poet.

One day an English teacher asked everyone in class to say what book they had most enjoyed during the past semester. She expected us to name such novels as *Ivanhoe* or *Vanity Fair*, which had been assigned reading, but when it came to my turn, I said, *The Adventures of Sherlock Holmes*. Everyone in the class tittered except the teacher, who looked pained.

I've been an ardent Sherlockian ever since Holmes first greeted Watson with those immortal words, "You've been in Afghanistan, I perceive." How as fine a writer as Doyle could have written an entire book about the reality of fairies is something I'll never understand even though I once wrote about it in an essay, "The Irrelevance of Conan Doyle." I was recently amused by a letter in the *Skeptical Inquirer* in which the writer pointed out that Doyle was not totally credulous about fairies. He wrote that although he was totally convinced that human fairies existed, he was not so sure about fairy horses and dogs!

A geometry class was skillfully taught by Miss Pauline Baker. My first book of *Scientific American* columns is dedicated "In memory of Pauline Baker, my first guide in the endless labyrinth." Before I left high school, Miss Baker astonished us all by marrying the school's basketball coach!

Alas, Pauline had her blind spots. I have told elsewhere about a day during study period when I was trying to determine which side has the win in tic-tac-toe. Is it the first or the second player, or is it a tie if both sides make their best moves? (Answer: it's a tie.) Pauline snatched away the sheet on which I was scribbling and said sternly, "When you're in my class, I expect you to work on mathematics and nothing else!"

I wrote lots of mediocre poetry when I was in high school, most of it published in the school's weekly newspaper, and later in a booklet titled *Poems: 1929–1931*, published by the school. I may be the only living person who preserved a copy. It contained my sonnet "Destiny" which goes like this:

> A leaf swayed with the breeze, then fluttering fell
> In silence to the whitened slope below.
> This and nothing more. Yet who can tell,
> As solitary watchers of the show,
> Who played the lead? Did fate or chance compel
> The fall? What circumstances long ago
> Then did the future of the act foretell,
> And plot the spiral pathway to the snow?
> So with life; tides in affairs of men
> Are simply outbursts of a pondrous chain
> Of countless forces brewing through the age.
> The rise of empires; fruits of sword and pen
> Converge and bring to focus deeds as vain
> And trivial as the turning of a page.

Years later, partly the result of reading William James, I became convinced that we can make freewill decisions that can change history in ways unpredictable, even by a God. Moreover, quantum mechanics has destroyed the notion of strict determinism. Imagine an airplane speeding across Europe with a nuclear bomb that is released by a click in a Geiger counter.

According to quantum theory, the timing of such clicks can alter history in ways that are unpredictable in principle, such as deciding whether city A or city B is destroyed. This simple thought experiment proves that, contrary to my sonnet, historical determinism is shattered by the laws of quantum mechanics.

An even rarer publication, *Essays in Prose and Verse*, published by the high school in May 1932, contains seven of my poems. I will here quote a short lyric called "The Pause" because it raises a profound question about which I actually once wrote a *Scientific American* article. Is it meaningful to assume it possible that time might come to a total stop, then, after a short or long pause, start up again?

Once upon a time (who knoweth when)
Time fell asleep and rested
For a thousand years.
And no one dreamed that while the speaker paused
To emphasize a point,
Eons came and fled.

Another poem in the same issue reflected my firm belief at the time that we are not alone in the universe. The influence of Lord Dunsany, whom I was avidly then reading, is obvious.

AN ETHRALDRIAN GAZES AT THE EARTH

Once upon a tiny planet,
Hidden in the depths of the milky way,
A creature stepped forth from his dwelling
Into the night, and gazed at the stars,
The soft violet rays of the moons,
Palely lighting his features.

His eyes sought out a single orb,
Far away in the constellation of Oth;
A gleaming point in the darkness,
And he wondered if it be a world,
Spinning through space like his own;
Bearing life and civilization
That wax and wane as the phases
Of Themor, the largest and brightest
Of Ethraldria's moons.
And he wondered if its inhabitants
Worshipped Amir the father,
Or Zada the god of the Humu's.

He continued to gaze and the planet
Continued to twinkle and scintillate
Against a background of infinite space.

The air was fresh—and he stretched his arms upward,
Marvelling at the glory and splendor of the universe.
Then turning, he re-entered his dwelling.

There is a whopping flaw in my poem. Unless the creature has incredibly powerful telescopic eyes, there is no way he could see the earth, though he might be able to see our sun as a dim star.

Many years later, influenced by Lord Dunsany's fine story "The Exiles Club" (in his *Last Book of Wonder*), I wrote a sonnet about the forgotten Greek gods. I sneaked a version of this poem into my *Visitors from Oz* novel, where it is recited by none other than Apollo!

Forgotten gods! Alas, the words convey
Too well the dreadful reason for our flight.
No angel host has fallen from the height
From whence we fell. Great gods who held the sway

Of kings and empires now are but the play
Of scholars. Altars once so warm and bright
With sacrificial blood beneath the light
Of ancient moons, lie crumbling, cold and gray.

Yet far beyond the vast Olympian snows,
In Oz we gods of Greece are living still.
And Jove in drowsy indolence still nods
His shaggy head in silence. Our repose
Is deep and calm, unbroken by the chill
Of disbelief. For who can kill the gods?

Near the end of every school year tests were given in all fields to decide who would become members of the prestigious T-Club. I became a member by making the highest score on a test in mathematics. When I reported for the meeting of new T-Club members, a teacher told me I had been selected to be the chairman of the peanut-sacking committee. This was a group of students who sacked peanuts for sale at next day's basketball game. After I finished supervising the sacking of hundreds of bags, I told the surprised teacher I was resigning from the T-Club. She couldn't imagine why!

My physics class was also a pleasure. It was taught by a Mr. Hurst. I later dedicated *Science Puzzlers* "To M. E. Hurst, a physics teacher who taught much more than physics." The "much more" was skepticism about the Bible. This, of course, was outside the classroom. We became friends. When the great Robert Millikan was scheduled to lecture at Oklahoma University, not far away at Norman, I rode with Hurst to Norman to hear the lecture. Of course I never suspected I would end my days in Norman.

I recall a gathering at Hurst's home where I met a friend of his who was active in Tulsa's First Unitarian Church. I was much amused when he told me that the most significant statement in the Bible was Pilate's question to Jesus, "What is Truth?" It could be the title of a book on the history of pragmatism. I later read William James's book *The Meaning of Truth*, which struck me as one of the funniest philosophy books I ever read. James struggled to clarify the pragmatic theory of truth without ever making it clear.

It was during my first year in high school that I read Thomas Paine's classic *Age of Reason*. Although Paine, like so many of the founding fathers, was a deist who believed in both God and an afterlife, his attack on the Bible as the word of God pushed me into a simple-minded atheism. Of course I never mentioned this to my parents, who attended Boston Avenue Methodist Church, but I recall that in high school assemblies, when someone led everyone in prayer, I proudly kept my head erect and my eyes open. During my final high school year, under the influence of an Irish camp counselor and Sunday school teacher at the First Presbyterian Church, I became a convert to a crude form of Protestant fundamentalism. I will cover this in a later chapter.

# 4

## CENTRAL HIGH, II

DURING MY HIGH SCHOOL YEARS I HAD TWO HOBBIES, chess and magic. Tulsa then had a room in the downtown Cole Building where every day, including Sundays, chess players would congregate. Mr. Cole, the building's owner, was a top chess player, probably on the master level. Every Saturday I would take a bus to the Cole Building, where I always found someone willing to play a friendly game. Of course I never played Mr. Cole or any other "heavyweights," as Cole called them. They included a lawyer named Neff and a salesman named Higgenbotham, and later a teenager called Bright Roddy, son of a beautiful Indian mother and a white father. Later Bright became chess champion of Oklahoma.

I never won a match with Bright. One day he taught me a game he had invented called wrestling chess. It starts with opposing kings at diagonally opposite corners of a chessboard. Players take turns jabbing their king with a finger until the two kings confront each other near the board's center. You then try to knock down your opponent's king in such a way that it is prostrate, with your king on top, pinning the other king to the "floor." Whenever your king is on its side, you can straighten it upright with a finger at its base.

I greatly enjoyed the chess games I played in the Cole Building. I remember a minister who would mutter, "What must I do to be saved?" whenever he was in a tight spot. A high school music instructor had the reputation of never losing a game because when he began to lose, he would take longer and longer to make a move, finally glancing at his watch and saying he was sorry to leave a "most interesting" game. Another player, the father of a girl in my homeroom class, had an eccentric chess opening he always played. He would advance his rook pawns, then move the rooks out and sideways to the center of the board! He often defeated much better players because he knew all the traps of his crazy opening, and his opponents did not.

Polish-born Sammy Reshevsky, one of the great chess grand masters (he once drew a match with Bobby Fischer) came to Tulsa to play simultaneously as many opponents as were willing to pay a small fee. Sammy won every game including one with me. Years later, when I was an undergraduate at the University of Chicago, I had the privilege of playing another game with Sammy. He was then enrolled in the university's business school, working for a degree in accounting. To the frustration of all of us who liked to meet on the second floor of the Reynolds Club, where there were chess tables, Sammy spent his spare time playing Ping-Pong in the building's basement.

We thought of a plan to get Sammy to the second floor. We knew he was poor, so we collected a small sum for a prize to go to the winner of a round-robin tournament. We posted details about the match on the basement's bulletin board, including the size of the prize. It worked. Sammy signed up for the match, and we each had a chance to play against him. Of course he won every game. I noticed that in playing against a potzer such as myself, Sammy made no effort to win quickly. Instead he spent his first dozen or so moves playing for position; then came the steamroller.

Sammy's appearance in Tulsa was his second visit to the city. Years earlier, as a child prodigy of age ten, he played simultaneously against a large number of opponents. A newspaper account of the event, tacked on the wall of the chess room, reported that the wife of one of the players asked Sammy if she could give him a kiss. "No," Sammy replied, "but you can kiss my manager."

It was a remark by Bertrand Russell that persuaded me to stop playing chess. He said somewhere that chess had become such an addictive time waster that he vowed to stop playing until he taught the game to one of his children. I made a similar vow. I did not play another game until I taught chess to my first son, Jimmy, and later to my second son, Tom. Since then the only chess I have played is with Gwen Roberts, a California high school math teacher who visits me occasionally at my apartment in Windsor Gardens. It's an assisted-living facility in Norman, where my son Jim is a professor of special education in the educational psychology department at nearby University of Oklahoma. Gwen usually wins.

My Apple computer has a good chess program. After wasting hours trying to beat it, I decided to stop playing as soon as I won a game. This finally happened. I must add that I won with a subtle mate in three that the computer, playing (I confess) on its lowest level, failed to foresee. I have not played chess since. One of my great difficulties in facing Apple is that it springs on me strange openings about which I am ignorant of the book moves.

My other high school hobby was magic. My father had taught me a few tricks, notably one involving bits of paper stuck on the sides of a table knife, and using what magicians call the "paddle move." Another of Dad's tricks that fooled me was the vanish of a wooden match from under a handkerchief. When I saw in a magazine an ad for the *Tarbell Course in Magic*, I persuaded my parents to subscribe. I still recall the eagerness with which I looked forward to each weekly lesson.

Later, when I lived in Chicago, I got to know "Dr." Tarbell. The "doctor" was for a degree in naprapathy, a crank medical practice like chiropractic, only worse. It was founded by chiropractor Oakley Smith (1880–1967) in Chicago, in the early 1900s. It still has a following in the United States and in parts of Europe, notably Sweden. In his youth Tarbell wrote and illustrated a book on physiognomy, a pseudoscience about how to read character from the shape of a person's nose, mouth, chin, eyebrows, and so on.

I had two good friends in Tulsa who were part-time conjurors. The older, Logan Waite, who owned a small manufacturing company, performed at local events. Working for him was a young juggler and magic buff, Roger Montandon. Roger ran a small mail-order business that sold juggling and magic equipment. He founded the nation's first juggling society, and the *Juggler's Bulletin*, the first periodical devoted entirely to juggling. The society, now called the International

Jugglers' Association, publishes *Juggler*, a handsome quarterly. Roger also sold *Cherchez la femme*, a cardboard puzzle I devised. Made of cardboard, the puzzle was to find how to fold the thing so it would reveal the picture of a nude woman. Mathematicians call its curious structure a tetraflexagon.

The most whimsical item Roger sold was called Sniff. It consisted of a small glass bottle, open at the top, with a tiny hook at the rim. Inside was a piece of rope. The idea was to ignite the rope, then secretly hook the bottle on a victim's back. He would go around the house smelling smoke but never able to locate its source!

Each year Tulsa Central High produced a vaudeville show called the *High School Daze*. Roger performed a juggling act on one of the shows. I too was onstage doing an act with friend John Bell. John's son Jeffrey, many years later, became an active conservative Republican who served as an aide for Reagan, Nixon, and Kemp. In 1938 he ran unsuccessfully for U.S. senator from New Jersey.

The act with Bell involved our appearance as a dwarf standing on a table. John's arms were in pants and shoes that looked like the dwarf's legs and feet. I stood behind Bell with my arms in coat sleeves to make the dwarf's arms and hands. I would wave the arms while Bell danced with his "legs." A decade later my sister, Judith, and a friend did a similar act for a *Daze* show.

My best friend in high school, who continued to be a good friend afterward, was John Shaw. Shaw did not attend Central High. A devout Catholic, he attended a Catholic school, but somehow we met. John was plump and I skinny. He liked to describe us together by holding up a fist with a finger alongside it to indicate me.

John had a great sense of humor that prompted a continual stream of funny remarks. He also enjoyed harmless practical jokes. One day, when I was riding in a car with John driving, he stopped by a curb to buy a copy of a local paper from a young boy who was hawking them,. Shaw tore the paper in half and handed a half back to the bewildered lad. "The paper is for my grandmother," Shaw explained. "She's blind in her left eye."

Shaw invented a game we liked to play when there was nothing better to do. When one of us drove the family car (in those days no high school student owned a car), the other would lower his head and keep his eyes closed while the driver took a circuitous route to a spot where he parked. The passenger then opened his eyes and tried to guess where we were.

As an adult, Shaw became the owner of Tulsa's leading bookstore, and an avid collector of rare books. I wrote an article about him titled "Tulsa's Fabulous Book Man" that ran in a local magazine called the *Tulsan*. (The magazine later published a similar article on Logan Waite.) Shaw's first major collection then was of the writings of Gilbert Chesterton, a British author we both admired. Shaw later gave his Chesterton collection, one of the world's largest, to Notre Dame's library. If you are puzzled over why I, a non-Catholic, can so admire G. K., I urge you to check my book *The Fantastic Fiction of Gilbert Chesterton*.

After disposing of his Chesterton collection, Shaw turned his attention to Sherlock Holmes and in a few years had one of the largest Sherlock collections in America. Shaw was an enthusiastic member of the Baker Street Irregulars, and whenever anyone asked him if Sherlock was real or imaginary, Shaw always replied, "Yes."

At one of the annual meetings of the Irregulars, Shaw brought down the house with an after-dinner speech on pornography in the canon. It is the only speech at an Irregulars meeting they were unable to publish in their journal. However, the speech *was* printed in a small fan magazine called *Shades of Sherlock* (vol. 18, August 1971).

Shaw opened his talk by saying he had recently decided to move from Tulsa to Santa Fe. To cut down the size of his library he decided to reread a book of short stories about Holmes to decide if he should keep all the books of the canon. He was shocked, he said, to discover that the book was saturated with porn! This will never do for my daughter, he went on, to pick up and read. Shaw then proceeded to cite examples of the porn. A character in a Holmes story says, "Last night I missed my rubber" (meaning of course a rubber of bridge). In another story Holmes casually remarks that he had that very morning "knocked up" Mrs. Hudson, his landlady. Shaw cited numerous references to men having "ejaculated," while his listeners rolled in the aisles.

Shaw enjoyed practical jokes of the sort that are funny to tell even though no one would even think of putting them in practice. For example, you board a crowded Manhattan bus dressed like a farmer and carrying a large bale of hay. You shove your way to the back of the bus; then after a few stops you get off the bus by way of its back door. A friend picks you up in his car and drives rapidly to the next bus stop, where you get on the bus again with the same bale of hay.

Shaw's finest example of a practical joke no one will ever do goes as follows. You remove the top of a toilet's flush box, sit on the opening, and relieve your bowels. You then close the cover and call your plumber. "Something is wrong with my toilet," you tell him. "Come as soon as possible." Shaw also always had a raft of great blue jokes, the products of what he liked to call a "clean dirty mind."

At times even a small practical joke can fall flat. Magic friend Paul Curry once told me about a time when a truck splashed mud over his jacket. The clerk at the cleaners asked his name when he handed her the soiled jacket.

"Sir Walter Raleigh," Paul replied.

This produced neither a laugh nor a smile. "I need your real name," she said.

I had a somewhat similar experience. I rolled my restaurant check into a tube and stuck it in my right ear. "I seem to have lost my check," I said to the cashier.

"Not funny," she said.

Shaw occasionally invented a word. One of his best was "obviosity." Someone once told me she had seen it in a dictionary where it was illustrated with a quote from something I had written. I no longer remember the name or date of the dictionary, and may be imagining I was ever told this.

In high school I was an ardent fan of Tulsa's baseball team, then called the Oilers. Uncle Owen, also a fan, took me to the games and explained all the little-known baseball rules. I can still recall and picture in my mind many of the players. There was Guy Sturdy on first, Flippin at shortstop, Lamb in center field. A pitcher named Black always grunted loudly when he threw the ball.

My brother, Jim, was an even greater baseball fan than I. We both enjoyed rereading "Casey at the Bat." My book *The Annotated Casey at the Bat*, a collection of sequels and parodies of Thayer's immortal ballad, is dedicated to brother Jim. I have often stated my firm belief that Thayer's ballad about the Mighty Casey will be read and memorized long after everyone has forgotten William Carlos Williams. You'll find in my *Casey* book a parody I wrote myself about the time Casey's son loses his pants on the way from third to home.

Baseball and tennis are the only sports I enjoy watching on TV. I'm not a fan of football. I dislike all sports that stress violence. It is difficult for me to imagine how anyone can enjoy a boxing match, the goal of which is for one man to knock another man unconscious. Wrestling has even more violence even if only simulated—a "ballet with violence," as I heard Minnesota's wrestler governor, Jesse Ventura, once call it. I find it equally hard to understand the popularity of bullfights in Spain that end with the poor bull's execution.

Tennis is the only sport I frequently played as a boy, because of the court in our yard. In high school I was on the tumbling and gymnastics team, and actually performed on the horizontal bar during a basketball game intermission! My father had responded to my request for a metal high bar in a side yard where I practiced almost daily. I didn't know then about the special gloves that gymnasts wear, so my hands quickly became covered with thick calluses.

During my one and only public performance my climax was swinging around the bar many times while hanging by my knees. I also did some handstands and rolls on the parallel bars. I never mastered back handsprings or backflips, but well into middle age I could turn forward handsprings on the grass. Watching young girls perform today on the bars is like watching miracles.

Throughout high school I suffered occasional spells of visual migraine. This is an ailment in which a headache is visually preceded by a blind spot that turns into zigzag scintillating lines that slowly move to the edge of the visual field. Lewis Carroll, a migraine sufferer, described the lines as "moving fortifications."

Puzzled by these visual effects, and fearing there was something wrong with my eyes, my folks sent me to an eye, nose, and throat man, who shall be nameless. He surely was a quack. He knew nothing about visual migraine and suggested my headaches could be linked to constipation. He advised me to take frequent mild doses of milk of magnesia! The same doctor treated me for sinusitis by poking up my nose a wire with a cotton tip saturated with red mercurochrome. It was not until I was an adult in the navy that I discovered the nature of my migraine, as I will tell in chapter 12.

Backtrack to magic. Although I have invented lots of tricks, mostly using cards, and written books on magic for the trade, I have never been a performer. I consider myself fortunate in this respect. Had I taken up conjuring as a profession (God forbid), I might never have become a writer.

The only time I was ever paid for doing magic was when I was in college and worked Christmas seasons in the toy department of Marshall Field's department store, where I demonstrated tricks from Gilbert's magic sets. I would gather a crowd, then perform magic using equipment from the largest set. It was then I learned that until you do a trick a hundred times for a live audience, you don't do it well.

One of my favorite impromptu tricks—a trick is impromptu if it uses ordinary objects and can be done at any time—is one in which you pretend to swallow a table knife. I once saw on a newsreel Douglas Fairbanks do this trick at a dinner. He did it well.

Fairbanks reminds me of a mediocre movie I saw as a child. Titled *Green Hell*, it starred Douglas Fairbanks, Jr. I mention it only because it contained a jungle scene in which Junior wrings out a washrag. I noticed he held the rag in a way different from how I held it. I always grabbed the cloth so my left and right fingers circled the cloth the *same* way. Junior held it with fingers curled the *opposite* way. Next day when I squeezed a rag I tried his grip and found it superior to mine. To this day, whenever I squeeze a washrag, I'm unable to avoid thinking of *Green Hell*. It's like being asked not to think of an elephant. I suppose such trivial compulsions are common, but it's the only case of a harmless syndrome I'm unable to get rid of.

I must have swallowed a table knife a hundred times. Magicians differ on how to end the trick. I always ended by taking the knife out of my left sleeve.

Onlookers assume that somehow, they don't know how, I managed to get the knife up a sleeve. That's not the secret, and spectators are just as mystified.

Friend Persi Diaconis, Harvard's eminent statistician, once told me that he had obtained an early French book on conjuring that described the knife trick exactly as done today. The author suggested you end it by standing up, bending over, and pretending to extract the knife from your rear end! He cautioned the reader to make sure you do this only for audiences you know would not be offended.

# 5

## HUTCHINS AND ADLER

Hutchins and Adler
Had careers of great promise
Before both were shot down
By the books of St. Thomas.

*—Armand T. Ringer*

CHICAGO! DEAR OLD LOOPY, AS CHRISTOPHER MORLEY called it in a little book of praise for the Windy City.

Having lived in Chicago for some fifteen years, I got to know the city well, on foot and with the help of streetcars and the elevated. It was my first introduction to a giant metropolis. Years later, when I lived a comparable time in Manhattan, I never felt the same about the city. It was too much like Chicago, but less friendly. Chicago is spread out. New York City, squeezed on a small island, is jammed upward. Tables in restaurants, and seats at counters, are far apart in Chicago. In New York they are close together.

New Yorkers are in a hurry. Chicagoans move slowly. They are more polite. They say, "Thank you" and "You're welcome." You can have a heart attack in Manhattan and lie on the sidewalk for thirty minutes before anyone notices you or phones 911. Even doctors won't stop to help for fear of a lawsuit.

I loved Chicago. Although I never hated Manhattan, the feeling was never quite the same. I had no urge to explore the city. I already knew what big cities were like. I read E. B. White's hymn to New York, which ended with his writing that "not to look upon" the city "would be like death." I said to myself, how can anyone be *that* enamored of *any* big city?

I lived in Chicago because the University of Chicago was there. I had intended to go to Caltech to become a physicist, but Caltech then required that students first spend two years at a liberal arts college. I gathered promotional literature from several universities that I assumed would accept me, and the literature from Chicago impressed me the most. The handsome Robert Hutchins, former dean of Yale's law school, had just been made president. At age thirty he was the youngest president of any major American college. At Chicago he had adopted what was called the New Plan, devised earlier by several professors. It was a radical change. Class attendance, for example, was never checked. You could "audit" any course without getting credit. However, you were required to take four survey courses on physical science, biological science, social science, and the humanities. If you passed a test for any of the four, you could skip it. This was true of other courses. By passing tests you could advance rapidly and obtain a bachelor's degree in a year.

These incredible freedoms greatly appealed to me after my poor grades in high school. As a freshman, sitting the first day in a class on English literature, I heard the instructor, Norman Maclean (he later wrote a best-selling novel titled *A River Runs through It*) say to us, "You never learned anything in high school. Now you are going to begin your education." The remark sent shivers along my spine.

I had no difficulty being accepted by the university. I passed a test for the physical science survey course, which allowed me to skip the course but get credit. I reveled in the freedom to attend any class. I think that during my four undergraduate years I audited more courses than I took for credit.

Mortimer Jerome Adler, Hutchins's good friend, was there. Hutchins had made the mistake of appointing Adler a professor of philosophy without consulting any of the philosophy faculty. This so enraged most of the philosophers that they resigned from the university, crippling the philosophy department for years. Hutchins was forced to move Adler to the law school as the school's only philosopher.

During the time I was at the university, Hutchins and Adler energetically promoted the Great Books movement, a movement that had earlier started at Columbia. The idea was that a general education required an acquaintance with the greatest books of the Western world. Later the University of Chicago would publish a set of the Great Books edited by Adler. The university bought the *Encyclopaedia Britannica*, also to be coedited by Adler in a new fifteenth edition. It broke the set into two parts: the *Syntopicon*, a two-volume index of short articles, followed by the usual multivolume set of long articles by experts.

Adler was a peculiar fellow. Raised by Ortho-
dox Jewish parents, he became enthralled by neo-
Thomism, a Catholic movement based on the works
of Saint Thomas Aquinas, the greatest of the medieval scholastics. For various personal reasons, which
he never made clear, for most of his life Adler refused
to convert to Rome even though intellectually he
believed the church's doctrines. In 1935 he gave a
speech, recorded and released in mimeograph form,
which I have carefully preserved to this day. In it he
stated that *if* the Catholic Church is what it claims
to be, God's one true church, then it was justified in
executing heretics! Adler later, greatly embarrassed
by this speech, renounced it, but, alas, the speech was
recorded for posterity.

In his classes Adler worked hard to convince students that Aquinas's five proofs of the existence of
God are valid. Much later he would clash with the
French Thomist Jacques Maritain over his (Adler's)
growing skepticism about the proofs.

In my senior year, as a philosophy major, I had the
following letter published in the *New Republic* (December 13, 1940):

> The text of Mortimer Adler's recent paper, "God and the
> Professors" (to which Sidney Hook replied in the Octo-
> ber 28 issue of your magazine), has just been printed in
> full in the student newspaper of the University of Chi-
> cago, and I have just finished reading it.
>
> As a former graduate student in the positivistic-
> minded philosophy department of the University, and a
> present resident of the campus community, I would like
> to make a plea to the readers of *The New Republic*.
> *Pray for the conversion of Mr. Adler.*

Mr. Adler has stated many times that he intellectually accepts the doctrines of the Roman creed, but that he lacks the divine faith necessary for conversion and entrance into the Church. There is strong traditional precedent for such an attitude. Gilbert Chesterton, for example, wrote his *Orthodoxy*, one of the greatest of modern Catholic apologies, almost fifteen years before he joined the Church.

So let us unite in prayer for Mr. Adler. And on the date that he enters Rome, let academic circles proclaim a day of rejoicing and thanksgiving. For Mr. Adler's brilliant and exasperating rhetoric will at last have found a home; and out of the dialectic fog will emerge a shape definite enough to be recognized, and solid enough to be worthy of honorable combat.

A few days after the letter appeared, I was sitting with a lady friend in Reader's drugstore, on the south side of the Midway, enjoying coffee. Adler and a lady were sitting not far away. He was staring at me intently. I suspect his friend had said, "That young man sitting over there is the one who sent that letter to the *New Republic*." It took many decades for the prayers I suggested to be answered. Shortly before his death at age ninety-seven, Adler was baptized a Catholic. He had earlier joined the Anglican Church, of which his second wife was a devout member.

I once described Adler as a man doing a comic walk with one foot on the curb, the other on the street. In many ways he had a superb mind. He authored many books of which the best, in my opinion, was *Art and Prudence*. His ego was enormous. If you check the *Syn-*

*topicon*, you'll find his picture heading a biographical sketch longer than the sketches of Bertrand Russell and other contemporary philosophers, none of whom rated a picture.

I once heard Bertrand Russell and Adler debate. The topic was whether there are eternal standards in education. Adler argued that Russell surely believed in such standards because he wrote a book titled *Education and the Good Life*. Did not the term "good life" imply standards of goodness? Russell responded by saying that Adler had been misled by his book's American title. In England the book was called *On Education, Especially in Early Childhood*.

An indication of the great influence of Aquinas on Adler, and Adler's great influence on Hutchins, is that Gilbert Chesterton's book *Thomas Aquinas* (the great French historian of Christian philosophy, Etienne Gilson, called it the best book ever written on the saint) was reprinted, in its entirety, in one of the volumes in *The Great Ideas Today*, a series of books edited by Hutchins and Adler. Hutchins's promising career was severely damaged by his association with Adler, and Adler's equally promising career was demolished by his infatuation with Aquinas. A "Peeping Thomist," someone once called him.

When philosopher Richard Rorty was working for his master's degree at the University of Chicago (his thesis on Whitehead was supervised by Charles Hartshorne), he wrote in one of his many letters to his mother that there was a campus rumor that Hutchins didn't exist. He was "merely a Great Thought in the mind of Adler."

I will have more to say about Hutchins and Adler in later chapters. For more details about their curious friendship, see the first chapter, "The Strange Case of Robert Maynard Hutchins," in my book *Order and Surprise*.

# 6

## RICHARD McKEON

Richard Peter McKeon
Was addicted to the opinion
That all philosophical points of view
Are equally admirable and true.

*—Armand T. Ringer*

ROBERT HUTCHINS BROUGHT TO THE UNIVERSITY OF
Chicago a philosopher even more bizarre than Mortimer Adler. He was Richard Peter McKeon. At that
time McKeon was the leader of a so-called Aristotelian
School of literary criticism. Years later he became top
promoter of a philosophical movement called pluralism. It was something like what in anthropology is
called cultural relativism. A cultural relativist is not
allowed to say that culture A is superior or inferior to
culture B. Cultures are obviously different, and one
can describe the differences, but there are no standards for judging one culture superior to another.
Philosophical pluralists say something similar about
the great philosophical systems of the past. They
clearly differ, but you can't say one is closer to truth
than another.

"I think it can be shown," McKeon wrote somewhere, "that ideological agreement on one philosophy
by all mankind is neither possible nor, if it were possible, desirable."

You must not say, for example, that Plato is superior to Aristotle or vice versa. You can't say Spinoza's
central vision is more likely true than Descartes's,
or vice versa. You should not suppose that William
James's version of pragmatism is better or worse than
the pragmatism of, say, John Dewey or Richard Rorty.

Surprisingly, McKeon's pluralism is identical with
the pluralism defended by Adler in his first book,
*Dialectic* (1927). In that book Adler argued that philosophy should not be concerned with truth, but only
with the play of one system against another. "The aim
of philosophy," Adler concluded, "might almost be
described as the attempt to achieve an empty mind, a
mind free from any intellectual presuppositions, and
unhampered by one belief or another."

An empty mind! Is it possible, I wonder, that
McKeon's mind was empty of basic beliefs? It is hard
to think so. Ironically, it was McKeon who first introduced Adler to Aquinas by telling him about a twenty-
one-volume edition of the saint's works translated into
English. Adler called the effect of his reading volume
1 "cataclysmic."

In the second chapter of *Order and Surprise* I tell
of my efforts to find out if McKeon's mind was truly
empty. Did he, for example, after abandoning his
Catholic upbringing, retain a belief in God? I knew
he had been raised a Catholic by a Catholic father
and a Jewish mother. Unlike Adler, if you read all
of McKeon's published writings, you will never find

out what he believed about any fundamental philosophical question! When he taught Plato, he was a
Platonist. When he taught Aristotle, he was an Aristotelian. When he taught Spinoza or Hobbes, he was a
Spinozist or a Hobbesian.

I wrote to George Kimball Plochmann, a former
student of McKeon who in 1990 published *Richard
McKeon*, the first biography. Plochmann had not the
slightest idea whether McKeon was an atheist, theist,
or something in between.

In desperation I finally turned to Zahava, McKeon's
second wife after Muriel, his first wife, died. Zahava
was a Jewish convert to Rome who had obtained her
doctorate under McKeon. To my pleasure and surprise she phoned back and we talked for half an hour.

Did her husband believe in God? Like Plochmann,
Zahava didn't know! She guessed and hoped that just
before McKeon died, he had returned to the faith of
his childhood, because he did not object to last rites.
But she wasn't sure. McKeon, as far as I know, is the
only professor of philosophy who never let anyone,
not even his second wife, know what Paul Tillich liked
to call a person's "ultimate concern."

Let me turn now to the Great Books movement so
dear to the hearts of Hutchins and Adler, and also to
McKeon. I agree that a liberal education should be
based, at least in part, on familiarity with the greatest
books of the Western world. For many years the Great
Books movement flourished here with its weekly discussion groups meeting in cities all over the nation.
The *Encyclopaedia Britannica*, owned by the University
of Chicago, published a fifty-four-volume set called
*The Great Books of the Western World*. A later 1990 edition

tossed out some of the books and expanded the set to sixty volumes. You can get lists of all the titles by checking Wikipedia on the Great Books movement.

Of course no two scholars would agree on what books to include in such a set. Adler's biggest mistake was including a raft of science classics that were indeed enormous breakthroughs at the time, but are now hopelessly dated and much too technical for today's average reader. To be educated you surely don't need to wade through Aristotle's scientific treatises, or books by Ptolemy, Copernicus, Kepler, Galileo, Newton, Bohr, Einstein, Eddington, Heisenberg, and other giants of science. You can't learn anything about modern math by plowing through Euclid's *Elements of Geometry*. To understand something about science the best plan is to read a short history of science, and popular works on relativity and quantum mechanics.

One of Adler's whopping blunders was to include among the Great Books Freud's *Interpretation of Dreams*. Freudian psychoanalysis died a few decades ago. To almost every psychiatrist today under the age of sixty, Freud, though a fine writer, has become the very model of a crackpot. Whenever he said something significant, it was not original. William James, in his *Principles of Psychology*, written when Freud was a boy, discusses at length the role of the unconscious in mental illness. And where Freud was original, he spouted baloney. His book on dreams, with its elaborate and preposterous symbolism, belongs to a set called *Great Books of Bogus Science in the Western World*.

In the literary field, Hutchins and Adler selected *Little Dorrit* as the one book by Dickens. Who is Little

Dorrit? Why is Joyce's *Portrait of the Artist as a Young Man* in the set rather than Joyce's masterpiece, *Ulysses*? *Don Quixote*, *Moby Dick*, and *Huckleberry Finn* are rightly there, along with Milton, Dante, and Shakespeare. Where are the great romantic poets?

If I were asked to select the Great Books (and who would ask *me*?), I would have included Lewis Carroll's two *Alice* books. Maybe even L. Frank Baum's *The Wizard of Oz*? I actually believe, so help me, that an educated person today should know both of those books before reading, say, George Eliot's *Middlemarch*.

At any rate, the Great Books fad has now almost expired except for a few remnant readers. You can still buy all sixty volumes from old book dealers, but why do so when inexpensive paperback editions of the classics are readily available, and in better editions than those chosen for the set.

The two harshest critics of the Great Books movement were Dwight Macdonald, who blasted the set in a notorious 1952 *New Yorker* article, and Alex Beam, a *Boston Globe* columnist. Beam even wrote an entire book, *A Great Idea at the Time*, on the topic. Beam called the books "icons of invulnerability—32,000 pages of tiny, double-column eye-straining type."

Hutchins did his best to persuade the University of Chicago to adopt the Great Books as the core of undergraduate learning, but without success. The only college that actually did this was the small St. Johns College, now at Annapolis and Santa Fe, after it was taken over by several pals of Hutchins and Adler. Few living rooms today have a Great Books set there to provide what Hutchins once called "colorful furniture."

It is hard to believe, but Richard Rorty, one of the nation's leading pragmatists, was heavily influenced by McKeon's pluralism. Susan Sontag and Robert Coover, two distinguished American authors, were other enthusiastic students of McKeon. On the negative side, Robert Pirsig, in his popular book *Zen and the Art of Motorcycle Maintenance*, has a harsh portrait of McKeon, as an intellectual bully, who is never named but called Chairman of the Committee.

I find it sad that such a brilliant historian of philosophy, an expert on Aristotle and the medieval thinkers, would so soon have faded into near oblivion. *The Oxford Companion to Philosophy* (1995, 1,009 pages) has no entry on McKeon. Neither does *The Cambridge Dictionary of Philosophy* (1995, 882 pages). Mortimer Adler also failed to make either volume.

Rudolf Carnap, at the University of Chicago when McKeon and Adler were there, has a long entry in both volumes. (Adler once called him "Carnap the Sap.") There are shorter entries on Charles Morris and Charles Hartshorne.

# 7

## I LOSE MY FAITH

WHEN I FIRST ENTERED THE UNIVERSITY OF CHICAGO I was in the grip of a crude Protestant fundamentalism. It was partly the result of my admiration for a counselor at Camp Mishawaka, a camp in northern Minnesota where my parents sent me for several summers. His name was George Getgood. He was also a Sunday school teacher at Tulsa's First Presbyterian Church. His sister, more devout than he, was one of the church's missionaries. For many months I attended the church's Sunday morning services and Getgood's class, though without becoming a church member.

It was Getgood, I believe, who introduced me to literature published by the Moody Bible Institute, in Chicago. I was strongly moved by a book of sermons by Dwight L. Moody, America's most famous evangelist before the days of Billy Sunday. I recall being especially impressed by Moody's sermon on the blood of Jesus. It was the only part of Jesus's body, Moody said, that was left on Earth. Christ's blood, he argued, runs like a scarlet thread through both Testaments. It also drips through the pages of my religious novel, *The Flight of Peter Fromm*.

Moody could not tolerate long and boring prayers during his tent meetings. When one such prayer went on and on, Moody approached the podium and said, "While our good brother is praying, let us all stand and sing hymn number so and so."

There was, I now write with acute embarrassment, a short period when I was interested in Seventh-day Adventism, a fundamentalist sect still flourishing around the world. I had found in Tulsa's library a book by Adventist Carlyle B. Haynes titled *Our Times and Their Meaning*. It persuaded me that the Second Coming of Jesus was just around the corner, an opinion held also by Getgood and by Billy Graham. I remember a morning when Getgood told his Sunday school class that 1933 was a very probable year for Christ's return!

For a brief time I actually attended a local Adventist church. I remember one Saturday morning when I was startled by a foot-washing ceremony. Every member of the congregation had his or her feet washed by another member. Not being a member, and with not-so-clean feet, I happily escaped the ritual.

My interest in Seventh-day Adventism evaporated when I came across a rare red-covered 1919 book titled *Life of Mrs. E. G. White, Seventh-day Adventist Prophet: Her False Claims Refuted*, by D. M. Canright. A former Adventist minister, Canright persuaded me that Mrs. White, one of the sect's founders, was in part a charlatan. Hundreds of paragraphs in her books were copied word for word, without quote marks or credits, from books by other authors. Many decades later Walter Rea, another disenchanted Adventist minister, would publish a larger work titled *The White Lie*, crammed with more examples of Sister White's shameless pilfering. After his book was published, *Time* reported that dozens of Adventist ministers had left the church after reading Rea's carefully documented volume.

Another former Adventist elder (as their ministers are called) who lost his faith was William Samuel Sadler. After leaving the church, Sadler became the guru of a rival cult called the Urantia movement. (Urantia is the cult's name for the earth.) Its bible, *The Urantia Book*, is a mammoth (2,097 pages) collection of what are called the "papers" written by higher beings on other planets! The papers were channeled by a sleeping "contact personality" whose identity the cult has never disclosed. In my book *Urantia, the Great Cult Mystery*, I give strong evidence that the sleeper was Sadler's brother-in-law, Wilfred Custer Kellogg. Kellogg and his wife, like Sadler and his wife, were also ex-Adventists. *The Urantia Book* is saturated with Adventist doctrines, such as the soul sleeping until Resurrection Day, the annihilation of the wicked, and the role of guardian angels. In recent years Matthew Block, an ex-Urantian, has uncovered large portions of *The Urantia Book* papers that are cribbed from works by others with no hint of credits or quotation marks. It almost seems as if Sadler was doing his best to become another Ellen Gould White!

My wife, Charlotte, thought writing my book on the Urantia movement was a huge waste of time, and I sus-pect she was right. I was intrigued by *The Urantia Book* partly because of its Adventist origins, and partly because I was intrigued by Sadler. I had read his early book *The Mind at Mischief* when I was a boy, and was much impressed. I was especially intrigued by an afterword in which Sadler said that in his encounters with trance channelers there were only two he was unable to dismiss as frauds. He did not name them, but, as I learned later, one was Ellen White; the other was the sleeper who channeled the "higher-ups" of *The Urantia Book.*

Did my hatchet job on the Urantia movement cause any true believer to question the book's claims? I doubt it. I know of no Urantian who left the cult because of my book. Sometimes a book blasting a religious movement only hardens the faith of a believer. Benjamin Franklin, in his autobiography, writes that he became a deist after reading a book attacking deism!

Back to Chicago. It was in my later years at the university that I decided it would be hypocritical to call myself a Christian. In spite of my admiration for what seemed the basic teachings of Jesus, I no longer believed he was God incarnate. I never lost my faith in God or the possibility of an afterlife, the two central aspects of Jesus's teaching. It was the preposterous miracles of both Testaments that did the most damage to my faith.

I could not imagine the creator of the universe engaging in such droll antics as turning Lot's wife into salt or parting the Red Sea. I could not imagine a God willing to drown every man, woman, child, and their beloved pets except for Noah and his family. I have told elsewhere about a lunch with a Seventh-day Adventist and his wife, when I asked how he could worship a God capable of drowning innocent babies. He caught me off guard with his answer. God, he said, knowing the future, knew that those babies would grow up to become wicked men and women. I was tempted to stand up and shout, "Touché!" It was a thought that had never occurred to me.

The miracles of Jesus fared no better in my mind. It was hard to believe that God, in human form, would engage in such magic tricks as turning water into wine, or multiplying loaves of bread and fishes. Above all I found it hard to suppose God would send anyone to eternal (eternal!) torment merely because, as Paul warned, they did not believe Jesus rose bodily from his tomb.

Matthew's Gospel tells us that when Jesus died, there was a great earthquake. Rocks were rent, "and the graves were opened, and many bodies of the saints which slept arose, and came out of the graves after his resurrection, and went into the holy city, and appeared unto many" (Matthew 27:52 and 53).

In the unlikely case that you, dear reader, are a hard-nosed fundamentalist, do you really believe this? Do you think God gave the skeletons of the saints fresh bodies, then covered them with clothes? Did the saints remain resurrected, or did they go back into their graves as skeletons? And if you doubt that all this took place, how can you trust the accounts of Jesus coming back to life?

It was a course in Geology 101 that widened the cracks in my crumbling faith. In high school, during my flirtation with Adventism, I found in Tulsa's library a big textbook called *The New Geology* by Adventist George McCready Price. Price persuaded me, to the sorrow of my geologist father, that the earth was about ten thousand years old and fossils were stone forms of life that perished in the great flood. A minimum knowledge of geology from my 101 course convinced me that Price was an amiable crank. He had no training in geology. Today even most of the intelligent design advocates do not take seriously the flood theory of fossils. The Adventist church no longer reprints Price's worthless books. His *New Geology*, with its clever but crazy arguments, may be the last and greatest final effort to defend a young Earth and the flood theory of fossils. If you are interested, you can learn more about Price by checking chapter 12 in my *New Age: Notes of a Fringe Watcher*.

In college I found myself faced with the following dilemma. If evolution is true, as I came to believe, then Genesis is not. And if Genesis is false, how could I trust the accuracy of other biblical events? If Eve wasn't fabricated from Adam's rib—and what a droll insult to women that myth is!—then perhaps Jesus never raised Lazarus from the dead after his body had started to decompose.

During my first year in Chicago I was one of the founders of the Chicago Christian Fellowship, a small band of fundamentalists very much out of place at a secular university. There is a picture of this group in one of the university's yearbooks. I continued to attend its weekly gatherings even after my faith had started to waver. At one meeting I spoke about the truth of evolution, and how one could accept it without giving up faith in God and Christianity.

My girlfriend in those confused days was Marian Wagner. Her mother was a devout fundamentalist, and Marian too was a member of the fellowship. Together we edited *Comment*, the university's literary quarterly. Our relationship was entirely innocent, and after graduation we drifted apart. Many decades later I was visited by Marian's daughter Bonnie Bedelia, a beautiful and talented Hollywood actress who has starred in many films. Bonnie was researching her mother's early years. Marian wrote excellent poetry, some of it published in *Comment*. I later passed along to Bonnie a clipping my mother had saved from some unknown newspaper of a tribute in verse Marian had written about Thornton Wilder.

Wilder was another friend of Hutchins who had joined the University of Chicago staff at Hutchins's request. I attended two of his classes. One was on the writing of fiction. To my embarrassment, several of my amateurish short stories were read aloud to the class by Wilder, then subjected to withering criticism by Wilder and the students.

The other class was a highlight of my college years. Wilder titled it "Masterpieces of the Middle Ages and the Renaissance." We studied three books: *Don Quixote*, Dante's *Inferno*, and a play of Shakespeare's. Wilder was a stimulating lecturer, and his enthusiasm was catching, I am forever indebted to him for introducing me to Cervantes's great novel, and to Dante's immortal epic—two classics I might never have discovered on my own. Wilder, let me add, was a theist, the son of missionaries to China. I was astounded to hear Wilder say on what he called "digression day"—a day on which he responded to student questions—that he had not yet made up his mind on whether Jesus was uniquely divine or just a great religious teacher.

Wilder's best seller novel, *The Bridge of San Luis Rey*, is about a bridge in South America that collapses, killing a number of pedestrians. Father Juniper researches the lives of those who perished in an effort to determine whether their deaths could somehow have been planned by God. He reaches no final conclusion. I discuss all this in an essay titled "Thornton Wilder and the Problem of Providence" that appeared in a Kansas literary journal. I have not included it in any collection of essays and reviews because I did not think it worth reprinting, but it shows the influence on me of Wilder's novels and plays.

I got to know Wilder slightly and even published one of his short-short stories in *Comment*. Our paths crossed one sunny afternoon on the Midway. We sat on the grassy slope and chatted for a while about one thing and another. It was 1936 and Chesterton had just died. Wilder had not heard of his passing. I asked if he had read Chesterton's *Man Who Was Thursday*. He had not. We spoke about Karl Barth, whose sermons I had been reading, and about Wilhelm Pauck's newly published *Karl Barth, Prophet of a New Christianity?* Pauck was on the university's Divinity School faculty, and I had been auditing one of his courses.

Wilder admired Barth and had only praise for Pauck's book as the first book about Barth in English. Pauck, by the way, is in my novel *The Flight of Peter Fromm* under a different name. I was astonished when many years later Pauck wrote a two-volume biography of Paul Tillich. It was hard to imagine two theologians being further apart than Barth and Tillich. Tillich pretended to be a Protestant but actually was a pantheist who rejected all the central doctrines of Christianity.

I did not know then that Wilder was gay. Indeed, this was not widely known until he and the boxing champ Gene Tunney were much in the news about their romp together around Europe. Wilder somewhere recalls an occasion when a book they were reading fell into a pool and Tunney retrieved it by diving and coming up with the book between his teeth.

Wilder's comic novel *Heaven's My Destination* is about George Marvin Brush, a Protestant fundamentalist who, in spite of his primitive views, comes through as a likable chap. It had poor sales, and bland reviews, but I read it with enormous pleasure and frequent winces. Brush has much in common with Peter, in my novel *The Flight of Peter Fromm*, and with myself in the days before Peter and I lost our faith.

# 8

## CHICAGO, I

Hog Butcher for the World,
Tool Maker, Stacker of Wheat,
Player with Railroads and the
Nation's Freight Handler;
Stormy, husky, brawling,
City of the Big Shoulders . . .

*—Carl Sandburg*

DURING MY FRESHMAN YEAR AT THE UNIVERSITY OF Chicago I shared a room on Ellis Avenue with Merle Giles, a high school friend. Later I stayed for a short time at a dormitory across from Ellis, then moved to a dozen or so different rooming houses over the next four years, the addresses of which I no longer remember except one—that of the Homestead Hotel at 5610 Dorchester. It was an old mansion long since replaced by an apartment building.

There is a chapter about the hotel in my novel *The Flight of Peter Fromm*. Its rooms were labeled A to Z, with X, Y, and Z in the attic. I lived in room X at a rate of ten dollars a week. I used to say I was kept awake at night by old ladies falling down the stairs. One day a portion of my ceiling fell. Several days later two funny-looking little men with strong foreign accents came to patch my ceiling. They could barely speak English. I later learned they owned the hotel!

One of my fellow residents at the hotel was Ed Haskell. In his youthful vagabond days he traveled the highways with his guitar. One afternoon a wealthy lady gave him a lift. She was so charmed by Ed's good looks and country songs that she established a fund for his higher education—a fund that lasted throughout his life.

When I first met Ed, he was majoring in philosophy at the University of Chicago, and deeply immersed in "general semantics," a cult started and headed by super-egotist Count Alfred Korzybski. The count lived in a building near the campus, at 234 West Fifty-Sixth Street. Korzybski had selected the house because he liked its 23456 sequence.

At the time we both roomed at the Homestead Hotel, Haskell was finishing his only book, a novel titled *Lance*. He either gave me a copy or I bought one; I forget which. The book was self-published and, as far as I know, never reviewed. You can imagine my surprise years later, when, living in Manhattan, I spotted Ed reading a book in Columbia University's library. Apparently he had transferred his graduate studies to Columbia. I didn't interrupt his reading.

A second even greater surprise came many years later when I reviewed for the *New York Review of Books* the autobiography of Willard Van Orman Quine, a famous American philosopher. It turned out that Quine and Haskell had been good student friends when they attended Oberlin College. Not only were they friends, but Ed was Quine's longest and *best* friend. Quine writes that he wept only twice in his adult life. Once at his marriage ceremony, and later when he was prevented from joining Ed at one of their many vacations together because the "head rascal," as Quine's son liked to call Ed, was ill and dying.

Quine touches only briefly on Ed's strange compulsion to tumble for idiotic cults. After his enthusiasm for general semantics began to wane, he next became— hold on to your hat—a Moonie! It was Haskell who persuaded the Reverend Sun Moon, a billionaire who owns the conservative *Washington Times*, to sponsor conferences at which scientists and other thinkers were asked to participate. Moon, by the way, finally announced that he was indeed, as long suspected by followers, the Second Coming of Christ. Quine tells in his autobiography of attending one of Moon's symposiums, and seeing the great physicist Eugene Wigner leave his seat. He hoped Wigner was leaving the conference because he was disgusted by what the speaker was saying, but no—he was only going to the men's room.

For a while *ETC*, a quarterly journal devoted to general semantics, was edited by the Japanese American Sam Hayakawa, author of a popular book on general semantics. Later he and the count had a bitter falling-out. In the early 1950s I ran into Sam one hot summer night at a small jazz spot near the university. I asked him what caused the break. He answered, "Words." Later, Sam became a distinguished U.S. senator from California.

In 1958 a follower of the count, David Bourland, Jr., introduced the word "E-prime" (English-prime) for a language that omits all variants of "to be," such as *is, am, are, was, were,* and so on. Using E-prime was supposed to add clarity to the language. For example, an old jingle could be translated into E-prime as follows:

Roses look red,
Violets look blue,
Honey tastes sweet,
As sweet as you.

It is hard now to believe, but for years a fierce controversy raged in general semantics circles over the value of E-prime. Papers and even entire books were written and published in E-prime! As far as I am aware, E-prime turned out to be useless, and today no one of any importance takes it seriously. Indeed, nobody of importance takes general semantics seriously. "Reading Korzybski extensively," wrote Noam Chomsky, "I couldn't find anything that was not either trivial or false." If you care to know more about E-prime, check chapter 5, "E-Prime: Getting Rid of Isness," in my book *Weird Water and Fuzzy Logic*.

There was a period of a year or two after I graduated when I had a job as social caseworker for the Chicago Relief Administration. This was during the final years of the Great Depression. I think my salary was less than what my clients were getting from the city. My case load was 140 families living in what in Chicago is called the city's Black Belt. My job was to visit clients regularly to check on how they were doing, and to make sure no one was secretly getting paid for a job. I doubt if many were, because in those days a job was hard to come by. One sad day I arranged for a casket and funeral of a man who had been taken care of at home by his wife because there were no vacancies at nearby hospitals.

Most of my clients were gentle, wonderful people. They would tell me about the "misery" in their arms and legs. Some of them who were Catholics had put together in a bedroom a small shrine complete with candles and a statue of Mary. One young man who lived on a fourth floor had arranged a string that ran from his room to the front door of the building. If you pushed a button beside his mailbox, he would pull on the string and unlatch the door.

Years later, married and living in Westchester County north of Manhattan, I rigged a similar string from the attic where I worked to the kitchen, where it rang a bell hanging over the kitchen door. My wife, Charlotte, would answer by pulling the cord to move a rattle on the wall by my desk.

Caseworkers were obliged after returning from an interview with a client to dictate the results of each visit. Dictaphone cylinders went to a room where their contents would be typed for administration records. One day I heard loud bursts of laughter coming from the dictation room. Someone had just typed a report in which I said that a client had told me he was suffering from hemorrhoids. I added, "Worker did not attempt verification." Miss Humphrey, a handsome black woman who was my supervisor, gave me a stern lecture on how I must take interviews more seriously.

Occasionally a client would be home but unwilling to answer a knock on the door. I had a trick to play. I would slide an envelope under the door, walk away with loud steps, then tiptoe back and plant a foot on the envelope's end. There would be a few vain tugs, the door would open, and I would say, "Hello, I'm your caseworker."

My introduction to the corruption of Cook County politics came when Merle Giles and I were asked by someone to serve as watchers while votes were counted in a dingy little office on Chicago's west side. A sweet-looking elderly lady recorded each vote by hand while Merle and I looked over her shoulders. She showed not the slightest interest in what names had been checked by the voter but simply called out and recorded the name of the politician favored by the Cook County machine. A policeman standing in a corner of the room came over to shut us up because we had been calling attention to what the lady was doing. He demanded to know whom we were working for. "Cook County Court," Merle shouted back. It did no good. We were forced either to keep quiet or be tossed out of the room. Afterward, I phoned the person who had hired us and told what happened. He thanked me for the report but said nothing else.

Followers of Hutchins and Adler were constantly slamming the social science faculty for lack of interest in philosophy. One afternoon the Adlerites sponsored a picnic at which was sung a fine parody on a university football song that begins, "Wave the flag for old Chicago. . . ." The parody went like this:

Wave the flag for social science.
They stand for facts alone.
Ever shall they be dogmatic.
John Dewey they enthrone.
With the pragamatists to lead them,
Without a thought they stand.
So cheer again for social science,
For they're zeroes every man.

One of the picnic leaders was Janet Kalven, a Jewish girl who became a Catholic nun under Adler's influence. Winston Ashley, who came from Blackwell, Oklahoma, became Father Benedict, a Dominican priest. Winston wrote poetry, some of which I published in *Comment*. The large number of students, both Jewish and otherwise, who became Catholic converts because of Adler gave rise to the saying that the University of Chicago was a Baptist college where Jewish professors were teaching Catholic theology to atheists.

The university had been founded by Baptist John D. Rockefeller. It was said that in the university's chapel, where the pastor had to be Baptist, the choir would sing, "Praise God from whom *oil* blessings flow." This reminds me of an anecdote about the college's first president, William Rainey Harper. Someone met him one day crossing the campus and asked where he was headed. "To the chapel," he replied, "where we intend to pray for a large donation."

"And you really think," Harper was asked, "that God will answer your prayers?"

"I do indeed," said Harper, patting his breast pocket. "I have the check right here."

I was sitting one day next to Winston Ashley, perhaps in Harper Library, when he suddenly burst into a laugh. He had been reading a book about the romance between Catholic philosopher Peter Abelard and Héloïse, one of his young students. A group of the girl's relatives, led by her uncle, Canon Fulbert, broke into Abelard's home and castrated him. Some Catholic historians have called this episode "beautiful." I consider it ugly and sordid. I have always been amused by Abelard's remark, somewhere in his writings, that "as I became more and more involved with Héloïse, I became less interested in philosophy."

Why had Winston laughed? He had just encountered a line in one of Héloïse's letters to her lover after she had become a nun. "Give all thou canst, and I will dream the rest."

In the original version of the song parody I quoted earlier, the fifth line is "With the grand old man to lead us." "The grand old man" was a phrase attached to Amos Alonzo Stagg, the university's famed football coach, before Hutchins, to the fury of trustees, abolished football. Football, said Hutchins, had nothing whatever to do with the purpose of a university, which is to give students a liberal education.

Hutchins had little love for sports. "Whenever I have an impulse to exercise," he once quipped, "I lie down until it passes away." He enjoyed other self-effacing remarks. After a divorce from his second wife, Maude, he said it was such an enjoyable experience he planned to do it more often.

Maude, by the way, was not only a prolific novelist; she also was an excellent artist. Her drawing of Hutchins graced the jacket of the president's book of speeches, *No Friendly Voice*. She collaborated with Adler on a bizarre little book titled *Diagramatics*. On its right-hand pages were sketches by Maude of nudes in various poses. On the left were paragraphs by Adler that made no sense but were supposed to imitate the style of various famous writers. If my memory is accurate, one of the paragraphs that began, "O blue art thou O last," was intended as a parody of Saint Augustine. What connection Adler's prose had with the opposite nudes beats me.

The two, Adler and Maude, actually gave a lecture about their book in Mandel Hall, which I attended. Adler read one of his paragraphs while Maude, on the other side of the stage, flashed on a screen one of her drawings. No one in the audience clapped. They looked totally puzzled. The book, in a limited and numbered edition, is now a collector's item.

Another Adler debate I attended, also in Mandel Hall, was with the biologist Anton J. Carlson. Adler, as usual, was neatly dressed in a tuxedo. Carlson came wearing a soiled laboratory apron. I forget what the debate was about, but from the applause Carlson seemed the crowd's choice of winner.

Both Hutchins and Adler were intensely disliked by most of the faculty. Someone once introduced me to the madman theory of education. It says it is good for a university to have a faculty member who is mad because opposition to his crazy opinions stimulates students into thinking seriously about fundamental questions. Adler was the University of Chicago's madman.

Although Hutchins was much less crazy, he tried to run the university like a dictator. One afternoon he sent a message to the famous physicist Arthur Holly Compton asking him to come at once to his office. Compton shot back a note saying that if the president wanted to see him, he should come to *his* (Compton's) office.

Some faculty members were so opposed to Hutchins that they wrote articles, and even books, attacking him. Harry Gideonse, a social scientist, published a small volume blasting Hutchins, and moved to another university to distance himself from Hutchins. Philosopher John Dewey was another Hutchins enemy who attacked him in print. The faculty heaved a huge sigh of relief when President Hutchins finally fled Chicago to run what he called the Center for the Study of Democratic Institutions, a California think tank. Amusingly, the center was funded by royalties from a best seller written by London-born Alex Comfort when he was at the center. Called *The Joy of Sex*, it was profusely illustrated with explicit art. Comfort later sued the center for not giving him what he considered his fair share of royalties. The center sued back, claiming Comfort had deliberately issued a much-revised version of his book, *More Joy of Sex*, just to lower sales of the first book. The conflict between Comfort and the center was a bitter one.

When World War II was getting underway, Hutchins, to everyone's astonishment, proved to be strongly opposed to America's entering the conflict. He gave a radio address that opened with this sentence: "America is about to commit suicide." On the campus Hutchins's chief opponent was none other than his friend Richard McKeon. McKeon followed Hutchins's radio speech with one of his own, strongly critical of Hutchins's isolationism.

Like McKeon, Hutchins was shy about making known his basic religious beliefs, though he made clear he was a theist. He was fond of saying there can be no brotherhood of man without a fatherhood of God, a favorite aphorism of liberal Protestant leaders. After one of Hutchins's speeches to students, in Mandel Hall's dining room, he took questions from the floor. I stood up to ask if he would mind revealing what his fundamental beliefs were. He said, "Yes," and pointed to another student for the next question.

One of the highlights of my Chicago days was having lunch with Vincent Starrett, then a columnist for the Chicago *Tribune*'s Sunday book review supplement. I greatly admired his pioneer study of the Sherlock Holmes canon, *The Private Life of Sherlock Holmes*. I also admired his poetry. Many years later I would quote his poem "Portent" ("Heavy, heavy—over thy head") in the introduction to my *Annotated Hunting of the Snark*. Still later, in my novel *Visitors from Oz*, I have the White Knight sing the following lyric. Starrett called it "Finally," but I prefer the title "Then."

> When you are tired of virtue
> And I am tired of sin,
> And nothing's left to hurt you,
> And nothing's left to win,
> Perhaps, greatly daring,
> Your eyes will question mine—
> But shall I then be caring
> For roses or for wine?

This is one of those rare short lyrics that are flawless in their blend of meaning and structure. Not a line can be improved. Vincent was pleased when I told him the poem could be sung to the tune "When Irish Eyes Are Smiling." Indeed, this is the tune the White Knight uses when he sings the song to Alice. I should explain that in my Oz fantasy, Wonderland and the huge chessboard behind the looking glass are actually underground in Oz, where they were visited by Alice in two out-of-body dreams.

Knowing my love of the Oz books, and my admiration of L. Frank Baum, Starrett gave me a copy of Baum's anonymous adult novel *The Last Egyptian*. He told me during lunch that a publisher had asked him to write the introduction to one of Edgar Wallace's endless mystery novels. Starrett asked what book he should select. I strongly recommended *The Green Archer*, which I consider Wallace's finest mystery. Starrett took my advice. I don't think he ever read the book, because his introduction never discusses the novel but is entirely about Wallace! For my fondness for this book, and my plot for a follow-up sequel, see my collection *Are Universes Thicker than Blackberries?*

Here is Starrett's best remembered poem:

221B

Here dwell together still two men of note
Who never lived and so can never die:
How very near they seem, yet how remote
That age before the world went all awry.
But still the game's afoot for those with ears
Attuned to catch the distant view-halloo:
England is England yet, for all our fears—
Only those things the heart believes are true.

A yellow fog swirls past the window-pane
As night descends upon this fabled street:
A lonely hansom splashes through the rain,
The ghostly gas lamps fail at twenty feet.
Here, though the world explode, these two survive,
And it is always eighteen ninety-five.

During the years I lived in the Windy City I led a curious sort of double life. There was a life centered on the university and friends I had there. My other life was in the world of magic, and centered on friends I saw only when I left the campus area and took the I.C.

(Illinois Central) to the Loop. All my Loopy friends were professional or amateur conjurors.

I had the pleasure of knowing all the Chicago magicians: Werner (Dorny) Dornfield, Johnny Platt, Paul Rosini, Eddie Marlo, Carl Ballantine, Bert Allerton, Paul LePaul, Jack Gwynne, and many others. They gathered regularly for lunch at a restaurant inside the Loop.

I remember one afternoon when Ballantine invited me to join him at a Loop theater that featured vaudeville. We saw Red Skelton do one of his rare magic acts. His great opening was to stride to the center of the stage, near the footlights, then fall into the orchestra pit! A highlight of his act was a performance with what magicians know as the "passe-passe bottles."

Red's comedy act was followed by a woman in a long gown who under a spotlight burst into an opera aria. Carl whispered to me, "She's wearing roller skates." Sure enough, when she finished singing and took a bow, she lifted her dress and skated off the stage, to return a few minutes later in a sensational roller-skating act! At that time Carl was beginning his acting career with a comedy magic act. Later he moved to Hollywood where he had numerous film roles.

The hangout for magicians in those days was Joe Berg's magic shop at the north end of the Loop. I ghostwrote Joe's *Here's New Magic*, a book long forgotten but containing some fine ideas. It includes a little-known way to tie a knot in a rope without seeming to let go of either end, a method superior to the old Hunter Knot. Joe's handkerchief or silk knot that gets tighter and tighter until it suddenly dissolves is a thing of beauty. How Joe thought of it beats me. I once had the privilege of showing it to the magician Cardini, at a gathering at Doc Daley's summer house in the Catskills. Cardini asked if the knot had a good "get-ready" (positioning/setting up the trick into an unusual position to make the trick work). I said no, but Dai Vernon was there, and he soon thought of a clever get-ready, which today I no longer remember.

Chicago's other magic shop was Laurie Ireland's, a walk-up outside the Loop on North Clark Street. Ireland had a whimsical sense of humor. One day I noticed an empty transparent box under the glass of a counter. Laurie had a card beside it that said, "The Invisible Pass."

An attractive young girl named Frances was hired by Laurie to clean the shop. It was not long until they were married. Frances took good care of Laurie in his final alcoholic days. Later she married magician Jay (Jasper) Marshall, and the two ran what is still called Magic, Inc., a mail-order house, from their warehouse and home on Chicago's north side. It was Jay's second and happier marriage. His former wife had been the daughter of magician Al Baker.

Frances wrote a delightful autobiography that had a great opening line, "Once I was seventeen." Jay, Frances, and Laurie were three much-admired friends.

# 9

## CHICAGO, II

DR. BEN REITMAN WAS GUEST SPEAKER AT A SOCIOLOGY class I audited—a class taught by Professor Ernest Watson Burgess. Ben began his talk by saying, "I don't want any of you to think I'm a Communist. I'm only an anarchist." All I remember now of what he said afterward was about his passionate love affair (it lasted a decade) with Emma Goldman, the most famous of female anarchists—that they broke up primarily because he wanted children and she didn't. Later Ben had a son and four daughters by several marriages.

I had two other contacts with Ben. One was when Burgess persuaded him to take the class on a tour, which I attended, of the flophouses on Chicago's decaying west side. The other was a visit I made, out of sheer curiosity, to an anarchist meeting on the city's north side. Ben was the speaker. Only about twenty persons showed up. It was a shabby, sad little gathering. After his speech Ben passed his hat. I tossed in a dollar bill.

The "Dr." in front of Ben's name was legitimate. He was a genuine M.D. and at the time I met him worked for Cook County as medical inspector of the city's prostitutes. He was known as the "hobo doctor," having spent most of his youth as a wandering bum. Ben published two unusual books, *The Second Oldest Profession*, and *Sisters of the Road*, a life of Boxcar Bertha. The oldest profession is prostitution. The second-oldest is pimping. Several biographies of Ben have been published. Emma writes at length about him in her 1951 autobiography, *Living My Life*. In 1972 *Boxcar Bertha*, a movie loosely based on the book, was directed by Martin Scorsese, starring Barbara Hershey and David Carradine.

Astronomy 101 was a course I took for credit, It was taught by Professor Edwin McMillan. He was an excellent teacher, but he had two prejudices. He believed women were intellectually inferior to men, and he was convinced that Einstein's theory of relativity was nonsense. He even wrote a book attacking the theory, the title of which I no longer recall.

Ronald Crane, of the English department, was second only to Richard McKeon as a leader of the Chicago Aristotelian school of criticism. I audited one of his classes that was devoted mainly to Fielding's novel *Tom Jones*. It prompted me to draw a caricature of Crane that appeared in the *Phoenix*, the campus humor magazine. Later I attended for credit a course on criticism taught by Crane and McKeon. The doctrines of the Aristotelian school were rather vague. As far as I know, the movement has since faded away.

Arthur Holly Compton, the famous Nobel prize winning–physicist, gave a lecture I attended during which he described one of his inventions, a clever device that proves the earth rotates and also identifies the direction north. The instrument was a large glass torus filled with water in which particles of matter are suspended. The tube is left standing until no motion of the particles can be detected by a microscope on the tube's side. If the tube is lined up on an east-west vertical plane, then suddenly flipped 180 degrees on its horizontal axis, the earth's rotation causes a slow drift of the particles. Their direction of flow indicates which side of the tube is north.

After Compton's lecture I approached him with a question. Suppose, I said, after flipping the tube on its horizontal axis, you turn it 180 degrees on its *vertical* axis, then follow with flips that alternate axis changes. Would this not increase the rate at which water flows through the tube? Compton took a half-dollar from a pocket and held it between thumb and finger while he rotated it along the two axes. Looking puzzled, he pocketed the coin and said he would have to think about it. My plan wouldn't work, but I was surprised the idea hadn't occurred to Compton.

Compton was a devout liberal Baptist. In one of his books he likened the afterlife to blowing out a candle, then igniting it again. At the time I considered this not a bad analogy, but it drew harsh scorn from secular humanists on the faculty.

Charles Hartshorne was a professor whose classes I took for credit. He was best known in philosophical circles for having coedited, with Paul Weiss, a set of volumes that pulled together almost all of Charles Peirce's published and unpublished papers. Peirce was a friend of William James, whose book *The Will to Believe* is dedicated to Peirce. Peirce was so furious with the way he felt James had distorted his term "pragmatism" that he changed it to "pragmaticism," a word he considered so ugly no one would steal it.

Hartshorne was some sort of pantheist, exactly what sort I never fully understood. He believed God was evolving in time, a dogma of the so-called process theologians, and philosophers such as Samuel Alexander and Alfred North Whitehead. God is limited in both power and knowledge. The future is not predetermined, partly because we possess free will, and partly because of randomness at the heart of quantum mechanics. Hartshorne believed that all life possessed some degree of consciousness, and that some animals, especially peacocks and bowerbirds, have a dim sense of aesthetics. He liked to bring birdsong recordings to his classes.

Hartshorne was a whimsical man who lived to be 103. Although he wrote many books, if he left any notable disciples, I haven't heard of them. Physicist Freeman Dyson has expressed admiration for Hartshorne's views, but how much of them he accepts I don't know.

I strongly disagreed with almost all of Hartshorne's opinions. For example, he believed that Peirce's notion of firstness, secondness, and thirdness was a monumental contribution to philosophy. I saw no good reason for stopping at thirdness. Why not a fourthness, fifthness, and so on? Hartshorne also belonged to a tiny band of thinkers who found Saint Anselm's ontological proof of God valid. The proof does indeed correctly maintain that an existing God is more perfect than a God who doesn't exist, but the leap from the *concept* of a perfect existing God to an *actual* one is so enormous that not even Thomas Aquinas considered it valid. Hartshorne wrote an entire book defending Anselm!

Hartshorne also held a strange belief about an after-life. For eternity, he said, we will always be the person who once lived and died, our existence firmly fixed in the mind of God. Apparently he took some sort of cold comfort from this view.

Another professor I got to know was Charles Morris. He was a great admirer of Peirce for his pioneer studies of semiotics, the theory of signs, a field in which Morris made significant contributions. He liked to draw diagrams with "sign" at the center of a triangle and the words "semantic," "linguistic," and "pragmatic" at the triangle's corners. Semantics referred to what the sign signified in the outside world, linguistics to the sign's relation to other signs, and pragmatics to the sign's influence on human behavior. Morris was also a great admirer of both Rudolf Carnap and John Dewey. He was able to persuade Carnap that there are no basic differences between the philosophies of the two men. I will have more to say about Carnap in a later chapter.

One of the pleasures of being a student at a great university is that famous persons come there to lecture. Two speeches that stand out in my memory were by Norman Thomas and by the writer Max Eastman. Thomas's lecture was actually a sermon delivered in the university's chapel. His topic was "The Right to Live." He reminded the congregation that every year millions of children, most of them in Africa, die of starvation. Modern agriculture is now so advanced that there is no good reason why it can't produce enough food to support the lives of every man, woman, and child on Earth. Yet somehow the world is unable to get the food adequately distributed to prevent millions of children from losing their right to live. It was a simple, powerful message, and I suspect the first time in the chapel's history that a congregation broke into applause.

Years later I heard Thomas speak at a debate in New York City. It was difficult for him to walk. As he slowly made his way to the podium, he uttered two words that brought down the house. He said, "Creeping socialism." As I type this chapter, democratic socialism is still creeping. In an effort to stimulate a dozing economy, facing a possible depression as a result of George W. Bush's refusal to regulate the stock market, greed took over, forcing a massive shift toward government intervention. "I overestimated the ability of the market to take care of itself," said Alan Greenspan in one of his rare intelligible remarks. It's hard to believe, but conservatives have only one remedy for a sick economy—lower taxes on the wealthy!

Right-wing Republicans are right in seeing Barack Obama's plan as creeping socialism. Will the United States manage somehow to return the economy to its previous relatively unfettered state? Time will tell.

The other lecture I vividly recall was Max Eastman's attack on modern poetry. Max first read some of his poems, notably "The Aquarium," which begins:

Serene the silver fishes glide,
Stern-lipped and pale, and wonder-eyed! . . .
They have no pathway where they go,
They flow like water to and fro.

Max then turned his attention to modern verse. He read what he said was a poem by E. E. Cummings, followed by some sentences taken from the ravings of an insane man. He asked for hands of those who found Cummings's poem the more pleasing. A few hands went up. Eastman then revealed that he had switched the two excerpts!

Years later I played on readers the same dirty trick. Under the pseudonym of George Groth I had written for the *New York Review of Books* a scathing attack on my just-published *The Whys of a Philosophical Scrivener*. I quoted what I said falsely was a lyric by William Carlos Williams called "The Red Wheelbarrow," then asked readers if they found it superior to some doggerel I had written. Like Eastman, I cruelly switched the two passages. "The Red Wheelbarrow" was my parody of a famous poem by Williams with the same title, and what I called doggerel actually were lines from a Williams poem. How successful my hoax was I have no way of knowing, but I suspect many readers tumbled for it. A friend later told me he had stopped reading my review in disgust, with a decision not to buy my book.

Thomas Vernor Smith, or T. V. as he was called, was a colorful philosophy professor from whom I took a course. He later became a U.S. congressman from Illinois and ended his days at Syracuse University. Smith loved to read poetry to his classes. One day he spent the hour on Carl Sandburg's *The People, Yes*. I was impressed enough to buy a copy of the book, which I reread every few years, always with pleasure. Sandburg, along with Stephen Crane, are two American poets I admire even though they wrote in free verse—a form of poetry I usually detest.

*Comment* sponsored and paid for a lecture by Carl Sandburg in Mandel Hall. Sandburg played his guitar and sang songs from his book *The American Songbag*. He then lived in a town south of Chicago. I occasionally spotted him riding the I.C. (Illinois Central) from his home to the Loop. Many years later my wife and I lived in Hendersonville, a suburb of Ashville, North Carolina, where Sandburg and his wife had settled on a goat farm.

One day T. V. Smith read to the class a lyric by Sara Teasdale. It so impressed me that I memorized it and still can repeat it word for word. It is a perfect poem. Not a word cries for change. Not a line seems contrived to fit the music:

> When I have ceased to break my wings
> Against the faultiness of things,
> And learned that compromises wait
> Behind each hardly opened gate,
> When I can look Life in the eyes,
> Grown calm and very coldly wise,
> Life will have given me the Truth,
> And taken in exchange—my youth.

Smith also read one or more poems by an obscure poet, Jamie Sexton Holme, who had been in one of his classes. In working on this autobiography I checked her name on Google and learned that three of her books were on sale at ridiculously low prices. I bought all three. *I Have Been a Pilgrim* (1935) is dedicated "for three fellow pilgrims" whose names are given only by initials. One of them is T.V.S., surely T. V. Smith. So he and Jamie must have been friends.

In another of her books, *Star Gatherer* (New York: Harold Vinal, 1926), the poem I liked best is "Spring Fever." It could have been a poem Smith read to his class. In any case, here it is. Critics will sneer at it because it has a pattern, but you don't have to listen to them.

> It's hard to be a good wife early in the spring,
> When the skies are so blue, and the butterflies a-wing.
>
> My own man's a good man, my children are dear,
> But there's just one moment at the turn of the year,
> With the buds all bursting and the
> green things growing,
> When I would be free as the four winds blowing!
>
> When I would go answering a strange bird's call,
> And not come back for anything at all,
> But run as fleet and far as a wild young deer
> To a green hill-side with a waterfall near.
>
> I'd sit all day on a warm white stone,
> And have a grand time with myself all alone!
>
> I shouldn't care at all who smoothed the feather-beds,
> Or put the dinner on to cook, or
> brushed the small heads.
> For I should hear the brown seeds
> and grass-blades talking,
> And see, among the aspens, a red deer walking.
>
> At dusk I'd hear the whisper of little creeping things,
> And in the boughs above me, the stir of bright wings.
> Close against the warm earth I would lay my ear,
> And she would have no secrets I would fail to hear.

I'd listen to the wee things stirring in the thicket,
And be friends just alike with the star and the cricket!

I wouldn't go home till the moon was overhead,
And when I got home, I wouldn't go to bed,
But sit on the door-step and look at the night—
For who can go to sleep when the moon is so bright?

My husband's a good man, my children are dear,
But it's hard to be a good wife in the spring of the year!

I quote the poem here because I think Jamie Holme,
like a thousand other humbler poets (as Longfellow
called them), deserves to be remembered. I know
nothing about Jamie except that she grew up in Mis-
sissippi and moved to Colorado with her husband,
Peter Holme.

After getting a bachelor's degree in philosophy, I
managed to snag a year's tuition from the Chicago
Theological Seminary, a campus school closely tied
to the university's Divinity School. The scholarship
came about through a recommendation by Professor
Wilhelm Pauck, a German theologian who had written the first book in English about Karl Barth. As I
reported in a previous chapter, it was titled with a
question mark: *Karl Barth, Prophet of a New Christianity?*
Along with a book on Barth's sermons, Pauck's book
came at a time when I was struggling to find a way
to save the remnants of a crumbling faith in Christianity. Barth was a not a fundamentalist, but he was
convinced the German Lutherans had strayed too far
from the heart of Luther's theology.

I found Barth inspiring and for a short time considered myself a Barthian. This was before Reinhold

Niebuhr persuaded me that the answer to Pauck's question was no. John Updike, by the way, went through a similar phase of infatuation with Barth. His comic novel *A Month of Sundays* is about a Barthian minister with a sexual problem. (See my review as reprinted in *Order and Surprise*.)

My novel *The Flight of Peter Fromm* contains a chapter on Pauck. I call him Von Cloven because he held (at least when I knew him) a belief similar to the heresy called Nestorianism. This is the belief that the human Jesus was quite distinct from the Christ of the Trinity. Later Pauck astonished me by writing a two-volume biography of Paul Tillich. I was surprised because I consider Tillich a humbug Christian. He once made the cover of *Time* where he was heralded as a great Protestant theologian, when in fact he believed in neither an afterlife nor a personal God, the two doctrines at the core of Christ's teachings.

Tillich had little use for the Jehovah of the Old Testament or the triune God of the New Testament. He defined God as Being, the "God behind the gods" of world religions. His theology was so close to atheism that Sidney Hook wrote a famous paper titled "The Atheism of Paul Tillich." Barth once called him a mere "journalist." Because no one denies that Being, or Existence, exists, the trick (as Hook observed) converts everyone by definition to Tillich's murky pantheism. Tillich's constant womanizing is covered in sordid detail in a book by his widow, Hannah. She tells how often she had to fire housekeepers because her husband was always trying to seduce them!

I once heard Tillich and Sidney Hook debate. Hook said that arguing with Tillich over his mushy philosophy was like punching a pillow. Tillich simply absorbed each punch by altering his shape. At one point Tillich waved his hand in Hook's direction and said, "There sits one of the most spiritual men I know." Hook, a dedicated atheist, reacted with a wry smile.

On another occasion I heard Tillich lecture at the University of Chicago's Divinity School. All I recall now is that afterward I shook Tillich's hand and told him I was taking a course from his friend Charles Hartshorne. Tillich corrected my pronunciation. In German, he said, the name is pronounced Harts-horn.

# 10

## I BECOME A JOURNALIST

AFTER A YEAR OF GRADUATE WORK I CONVINCED MYSELF
I had no desire ever to teach philosophy. I wanted to
be a writer. Seeing no point in aiming for a higher
degree, I left the university and returned to Tulsa.

My dad was a friend of an Irishman named Andy
Rowley, then oil editor of the *Tulsa Tribune.* When he
told my father he needed an assistant, my dad suggested me. After being interviewed by Rowley, I was
hired by the *Tribune* at a salary of fifteen dollars a
week. My main job was to visit daily all the oil offices
in Tulsa, then known as the "oil capital of the world,"
to check on the status of drilling wells. I would report
such stirring news as that Gulf Oil Company's well,
at such and such a location, had reached a depth of
so and so. On Sundays I provided a gossip column of
news picked up from conversations with persons I met
in various offices.

Occasionally I was assigned a feature story. For
example, I interviewed a young woman in a fishbowl. The Orpheum Theater was promoting a movie
by having in its lobby a large fishbowl in which a
scantily clad girl seemed to be sitting. I have no recol-
lection of what we talked about, and hope my interview was never preserved.

A more significant assignment was to interview employees at a newly opened office of a seismograph company. At that time the seismograph was slowly becoming an essential tool for oil exploration. Oil is always found under limestone domes. Underground water flushes oil upward to become trapped under the domes. A seismograph sets off explosions, then records the time it takes for bouncing sound waves to detect a dome. If a dome is found, there may or may not be oil beneath it. If no dome is found, there is no need to drill an exploratory well.

I got to know two employees of the seismograph firm. One was Tom Gilmartin, a physicist from England. For a while he and his wife and children lived in an apartment above my family's garage. After the Gilmartins moved to a house, the apartment was rented to another employee of the seismograph company, Bruno Pontecorvo, and his wife and children. Bruno was a physicist who had previously been an assistant to Enrico Fermi, in Italy. He seemed totally apolitical, never talking about politics or economics. After leaving Tulsa, he worked in several nuclear physics laboratories here and in Canada and England. In 1950 Bruno and his family abruptly left Rome, where they were on a holiday, and took a plane to Russia, where he was put in charge of Russia's nuclear energy program! There is no evidence Pontecorvo was ever a spy, but it turned out his brother had been active in the French Communist Party and Bruno had long held pro-Stalinist sympathies.

Before he settled in Russia, Bruno made significant contributions to nuclear science, notably his theory that neutrinos continually oscillate among three different forms. I remember how astonished Gilmartin was when he first learned of Bruno's defection.

Another friend who worked for the seismograph company was Jacob (Jake) Neufeld, a Polish mathematician who read and spoke Russian. He, too, was flabbergasted by Pontecorvo's defection. Jake later joined the laboratory at Oak Ridge, where his main job was translating Russian papers. He once provided me with translations of Russian articles attacking relativity and quantum mechanics as worthless bourgeois theories. It took a long time for Russian science, mired in Marxist ideology, to recognize the validity of both theories. This information from Jake enabled me to write an article titled "Bourgeois Idealism in Soviet Nuclear Physics." It appeared in the *Yale Review* and was later reprinted in my *Science: Good, Bad, and Bogus.*

I lasted only a year or two at the *Tulsa Tribune.* Its editor, Jenks Jones, son of the publisher Richard Lloyd Jones, was, like his father, a far-right conservative. One day in an elevator he saw me with a copy of the *Nation* in my hand. I suspect this was a reason he decided I had no future with the paper, and I was asked to leave. Although my period of employment was brief, I learned a lot from it. I can still picture the cluttered, smoke-filled floor, spittoons at every desk, no women in the room, and the smell of printer's ink drifting up from the basement where linotype operators were setting type.

With no prospect of another Tulsa job, I took a Greyhound bus back to Chicago* where, by a stroke

---

* The bus ride was in midwinter and the bus was unheated. As we pulled up to a rest stop, an elderly lady behind me, who had boarded the bus in Arkansas, said in a loud voice, "I'm going to order me a bowl of hot soup and put my feet in it."

of good fortune, I landed a job in the University of Chicago press relations office. It came about as a result of a friendship with Don Morris, then a press relations writer. When I edited *Comment*, the university's literary magazine, Don edited the *Phoenix*, the campus humor magazine. In addition to my caricatures of Charles Morris, Ronald Crane, and Mortimer Adler, the *Phoenix* also published some of my joke cartoons, and two pictures I drew for "The Ballad of Terrible Mike." This was a poem by friend Sam Hair, who decades later reprinted it as a pamphlet. Sam's ballad imitated "The Ballad of Yukon Jake," by Edward Paramore, Jr., a poem that *Vanity Fair* published no fewer than three times. You'll find Paramore's ballad in my *Famous Poems from Bygone Days*, along with some little-known facts about this one-poem author.

My job in press relations was mainly writing releases about science research going on at the university. It was a happy period of my life. One of my releases told about the discovery of mysterious fossil spirals called Mima mounds. The release so amused a *Chicago Tribune* rewrite man that he turned my piece into verse that was printed as prose. I imagine no reader ever realized that the account both rhymed and metered. Another release reported on the discovery of a fine specimen of fulgurite, or fossil lightning, that arced through an Illinois sand dune.

I learned a lot about writing releases from Don. Later he became editor of *Chicago*, the university's alumni magazine. Still later he joined the staff of *Life* magazine. When I compiled a selection of parodies of "The Night Before Christmas" for my book *The Annotated Night Before Christmas*, I persuaded Don to contribute a fine political parody.

Don died years before he should have of lung cancer. He had been a heavy smoker. Don was then living in D.C. with a job the nature of which I have forgotten. He left a manuscript for an unusual novel titled *Paper Trails*. It consisted entirely of letters, memos, diary entries, court records, newspaper clippings, and so on. I read the manuscript and thought it worth publishing. The manuscript now may be gathering dust in a relative's attic.

At desks near me in the press relations office were two other friends. Brownlee Haydon was the son of Albert Eustace Haydon, chairman of the university's department of comparative religion and the model for Homer Wilson, the narrator of my novel *The Flight of Peter Fromm*. In this wild novel Homer is partial to fancy vests that symbolize his fondness for the great religions of the world, none of them true. Haydon was a firm secular humanist, a fancy term for atheist. His son Brownlee became a staff writer and researcher at the Rand Corporation.

Cody Pfanstiehl, the other office friend, became director of the Washington, D.C., transit system. I recall him bringing down the house on an occasion I have forgotten by singing a song he had composed about the works of a sociologist that distinguished between the upper upper class, the upper middle class, and the upper lower class; and the middle uppers, the middle middlers, and the middle lowers; and the lower uppers, the lower middlers, and the lower lowers.

Cody invited me one time to attend a summer camp for girls where he played a fiddle for a square dance. The dance was followed by a campfire at which Cody played a solo on his violin. After the solo, which I found moving, I asked him the name of the piece he had played. There was no piece. He was improvising.

Cody was close to a saint. He had the highest of moral standards. After his first wife died, he married a blind woman. Both were active in promoting causes for the visually impaired. One of Cody's sons contributed a funny parody to my anthology *The Annotated Night Before Christmas*.

A fourth friend from my press relations days is Milton Semer, then a law school student. He came daily to the press relations office to run off mimeograph copies of our press releases. Milton became a D.C. attorney, active in the nation's politics. We have kept in touch over the decades and talk often by phone. He is my source for inside information about the skullduggery that goes on behind closed doors in our nation's capital.

Another classmate of that period with whom I kept in touch over the years was David Eisendrath. He became one of New York City's top photographers. His great photo of Einstein, taken on the day he became a citizen, ran on the front page of a Manhattan newspaper. I asked Dave to enlarge it for me, and it now hangs framed over my desk. I love the picture. A puff of pipe smoke looks like a white goatee. You can see the tiny flag in Einstein's lapel, a pin given to all new citizens. Einstein is neatly dressed except, I was told, for an absence of socks.

Einstein, the most creative physicist since Newton, is another of my heroes. I wrote a book on relativity mainly to teach myself the theory. It is still in print as a Dover paperback titled *Relativity Simply Explained*. I would like to write a companion book about quantum mechanics, but I'm now too old to attempt it. Einstein admired the theory for its consistency and power, but thought it incomplete—"God does not play dice"—to be modified some day by a deeper theory that would eliminate fundamental randomness. He could be right.

When the University of Chicago celebrated its fiftieth anniversary, one of my tasks was to arrange an exhibit of all the works of fiction that had the university as a setting. They start with *Maroon Tales*, a collection of short stories by the humorist Will Cuppy, and include a surprising number of novels by later authors. The most notorious was *Grey Towers*, written anonymously by Zoe Flanigan, a young English instructor who had graduated at the university in 1911. The book was published in 1923 by Pascal Covici. It was rumored that Zoe married Covici, but I was never able to verify that.

On the plus side the novel outlines what would eventually be called the New Plan. It was put into effect by Hutchins but actually was designed by several University of Chicago professors. On the negative side, *Grey Towers* goes into lurid details about the adulterous sex lives of thinly disguised Chicago faculty members. Lawsuits took the novel out of circulation.

One of my Chicago friends was Tony Eidson, a former student who, like me, had majored in philosophy. Tony and I each called the other Jake. This whimsy started when we heard someone tell a joke about a salesman who asked the lady who answered the doorbell, "Madam, do you sleep in a nightgown or pajamas?"

"Neither," she replied. "I always sleep nude."

The salesman put down his suitcase, extended a hand, and said, "My name is Bowers, Jake Bowers."

The joke is almost meaningless shaggy-dog nonsense, but somehow it amused us.

Before I forget, let me say something about the Young Communist League. It was active on campus during the years I was there. An attractive girl with bright red hair and a great sense of humor was Virginia (Ginny) Miller. I dated her a few times. She was an active member of the league and I was then a sort of fellow traveler, fascinated by Russia without any comprehension of what was actually going on there. Well, one day I had the pleasure of introducing Ginny to Paul Goodman. Paul had been a student of McKeon at Columbia University and had followed him to Chicago.

Paul later became best known for his book *Growing Up Absurd*. He also won a large underground following with his novels, short stories, and poems. At the time I first met him, he was a great admirer of Wilhelm Reich and his orgone therapy. I can't recall now if he actually sat nude in one of Reich's orgone boxes to build up his orgone energy, or whether he recognized and dismissed Reich's obviously crank side. Paul also was having difficulty seeing. He had read a book on eye exercises—a book endorsed by Aldous Huxley—and had thrown away his glasses in an effort to cure nearsightedness by wiggling his eyeballs.

I hung out in those days in a back booth at the Maid-Rite Grill, on Fifty-Seventh Street, where for a time I worked as a waiter and soda jerk. Among those who came regularly to the booth was Frank Meyer, a Communist Party functionary in charge of the campus area. He was constantly trying to persuade me to join the Party. Whenever we met, he would always greet me with "How long, Martin, how long?" He meant how long would I delay becoming a comrade. Years later Frank joined the ranks of those who, like Whittaker Chambers, become disenchanted with Marxism and turn far right. Frank ended up as an editor for Bill Buckley's *National Review*. He commuted to New York City from Woodstock and, shortly before he died, became a Catholic convert! I recall phoning him one afternoon at the *National Review* office and suggesting we get together. I too had become totally disenchanted with Marx, and was writing book reviews and articles for the anti-Stalinist, democratic socialist weekly the *New Leader*. So Frank and I had something in common, but we never again met.

As usual I'm getting sidetracked. One day when Paul Goodman and Ginny were both at the Maid-Rite booth, I introduced them to each other. They fell in love and soon were married. After producing a daughter, they eventually divorced. Paul wrote a poem titled "Falling Out of Love." To this day I can't understand why Ginny married Paul. She must have known he was gay. At any rate, I ran into Ginny one day when she was back on the University of Chicago campus, and invited her to dinner. I recall asking if she had ever found out what Paul's fundamental philosophical beliefs were. For instance, did he believe in God? Ginny smiled and slowly shook her head. She had no idea what Paul believed. Did he continue to claim he was an anarchist? Ginny didn't know. Paul was still a good disciple of McKeon.

I have a dim memory of a political meeting of some sort sponsored by the Young Communist League. Ginny was the chairwoman. A young man stood up and delivered a long, tiresome, irrelevant comment, then sat down. Ginny's one-word response provoked howls of laughter. All she said was "Balls!"

Years later, when I lived in Greenwich Village, I ran into Ginny at a Village bar on Bedford Street called Chumley's that was popular then with young bohemians. She told me she was there every Friday night. That was the last time we met, although one afternoon I passed her when she was looking into the window of a Village bookstore. A small girl, she was as cute as ever, her red hair in a ponytail. I was then happily married and with a son. She never noticed me, and I walked by without speaking.

# 11

## MOTHER AND DAD

THE TIME HAS COME, IN THIS SLOVENLY AUTOBIOGRAPHY, to speak about my mother and father. My dad, James Henry Gardner, grew up on a farm in Sonora, Kentucky, a small town not far from Louisville. His father, for whom I am named, was a gentleman farmer. Chores were all done by hired hands. He had two sons, my dad and Uncle Emmett, about whom I have written in a previous chapter.

After graduating from the University of Kentucky, in Lexington, my father worked for a while for the U.S. Geological Survey, mapping rivers in the then stateless New Mexico, and later hunting fossils for the Smithsonian Institution. Eventually he obtained a doctorate in geology from George Washington University, in the nation's capital.

Dad's first move in the direction of the oil business was his discovery of a source for fuller's earth. Fuller's earth is porous clay so in want of water that if you touch your tongue to a piece, it is difficult to free the tongue. At that time the clay was essential to oil refineries as a filter for crude oil. Dad opened a mine in southern Illinois where he had discovered the clay, and for several years the mine was a source of steady income. When refineries found a better way to filter crude oil, the market for fuller's earth disappeared.

Dad turned his attention toward petroleum geology and the rapidly growing oil business. He and his new wife, a Lexington girl named Willie Spiers, moved to Tulsa where he formed the Gardner Petroleum Company. It later consisted of himself, an accountant named Harry White, and a secretary. Dad did all the oil prospecting. When he found a promising spot to drill, he would hire a drilling company, then send his good friend Lucien Walker into the area to lease land from nearby farmers. The test well would usually be dry, but now and then it would strike oil, and the Gardner Petroleum Company would prosper.

My parents were a sort of mixed marriage so common in America and throughout the world. My mother was a devout Methodist who believed the Bible was the Word of God, and Jesus was the Son of God, but exactly what she thought about the great biblical miracles I was never sure. Not once in my entire life did I question her in any detail about her beliefs, nor did I ever hear my parents discuss a doctrinal question.

My father, I later learned, was Christian in name only. He went regularly to church with my mother, but as far as I could tell was a pantheist who identified God with Nature. His love of nature was unbounded. On the shelves of our living room were complete sets of the writings of Henry Thoreau and John Burroughs. My dad greatly admired both men, although he considered Burroughs the better scientist. One day he told me about Thoreau's monumental blunder. In his diary Thoreau speaks of standing at the end of a rainbow that arched up and away from him! Dad read to me a passage in an essay by Burroughs in which he slams Thoreau for such an incredible falsehood. Did Thoreau, Burroughs asks, look for the pot of gold?

One afternoon, on a trip east, Dad took me to Thoreau's cabin in the Massachusetts woods. I picked up a smooth rounded gray rock, with white streaks, at the edge of Walden Pond. I still use it as a paperweight and occasionally as a hammer. Thoreau, my father said, might well have once held that very stone.

Dad's main hobby was bird-watching. He kept a journal in which he recorded the days on which he spotted visiting birds. He actually owned a huge elephant folio of Audubon prints that had been owned by Mark Twain and was even inscribed "Samuel Clemens." Each page, suitably framed, could have sold for a small fortune. Alas, Dad gave the book to a Tulsa museum and took a tax cut. The first Tulsa Audubon Society was organized by my father.

Two memories remain about Dad's whimsical observations. He once asked me if I ever noticed that on pages of books you often see paths of white space between words that wiggle down the page like worms. He wondered if a page ever had a white worm that ran from top to bottom, or would this be as rare as being dealt a bridge hand of all one suit? On another occasion, while we were on a train, he pointed out that if you closed your eyes, you could imagine the train speeding the opposite way, or even moving sideways. When sleeping on a train, he said, he liked to induce sleep by sending the train in different directions.

On a more serious side Dad was constantly teaching me elementary science. I was a child when he explained why a rainbow was round, and why no two people see the same bow. He explained why rays of sunlight you sometimes see spreading like a fan from a hole in the

clouds are actually parallel, like railroad tracks con-
verging in the distance. As a youngster I learned from
him how to locate the Big and Little Dippers, Orion,
Cassiopeia, and other constellations. I learned why
thunder comes a few moments after you see lightning
flash, and why thunder rumbles. He explained the
phases of the moon, and why compasses point north.
In brief, his love of science left a huge mark on my life,
for which I am deeply grateful.

I don't know if my mother believed in evolution,
but I do know my father did. I found among his
papers a clipping from a Tulsa paper about a speech
he gave at a Rotary Club luncheon in 1923. Its head-
line is "Story of Creation Told in the Bible Is a Contribution to Mythology, He Believes." Although Dad
praises the Bible as a great "rule and guide of life,"
and says Jesus was in some sense divine, he makes
clear that the earth is hundreds of millions of years
old, and the Genesis account of creation must be
replaced by a recognition of the slow evolution of
all forms of life. God, he said, is "universal love and
truth and goodness."

My father's second hobby was Oklahoma history.
A careful reading of Washington Irving's *Tour on the
Prairies* enabled him to locate the exact spot where
Irving and his men had once camped. It was a place in
Oklahoma where the Cimarron River and the Arkansas River join. He later paid Frank von der Lancken, a
Dutch artist living in Tulsa (his wife, Julia, taught art
at Central High), to paint a beautiful oil picture of the
two rivers, a painting now owned by my son Jim. Alas,
the scene no longer exists. The meeting of the two rivers is now the site of a large reservoir.

I recall an evening when there was a long discussion by guests at our house over whether the moon does or does not rotate on its axis as it circles the earth. Because the moon always keeps its same face toward the earth, some of the guests argued that the moon does not rotate. My dad agreed, or pretended to agree. He proposed the following thought experiment. Imagine, he said, a mammoth plank with one end attached to the earth, the other end nailed to the moon so it can't rotate. As the plank swings around the earth, it carries the fixed moon with it.

The debate, of course, is over how to define "rotate." Relative to the earth the moon does not rotate. Relative to a viewer on Mars, it rotates, once per each revolution. The question is similar to one posed by William James in the first chapter of *Pragmatism*. A hunter is aiming his gun at a squirrel on the far side of a tree trunk. As the hunter circles the tree, trying to shoot the squirrel, the squirrel scurries around the trunk, always facing the hunter. After the hunter and squirrel have both circled the tree, has the hunter gone around the squirrel?

Again the question is trivial because it depends only on the meaning of "go around." The hunter *has* gone around the spot where the squirrel is located, but he has *not* gone around the squirrel if "around" means facing all sides of the squirrel. It is hard to believe, but at one time the moon puzzle aroused so much controversy among readers of pre-1900 *Scientific American* that the magazine devoted a special issue to the debate!

Now for an interesting related puzzle that I will leave to the reader. If you were on the moon, you surely would see the earth rotate. Would you also see it move across the sky?

One day I complained so much about high school, which I considered a sheer waste of time, that my father allowed me to stay home for the entire day so I could finish some project I had in mind. To the school he wrote the following letter. "Please excuse Martin's absence yesterday. He was sick at home with the gripes." Of course the principal's office assumed my dad didn't know how to spell grippe!

I once persuaded Dad to buy a saxophone. I took a few lessons, but because I have such a poor ear for music, the most I ever learned was how to run a scale. Many years later I suggested Dad sell the instrument. His classified ad in a local paper said, "New saxophone for sale at fifty dollars or a whole lot less." As I recall, a young man showed up at the door and bought it for ten.

Many restaurants in Oklahoma, when I was a boy, had slot machines. Once on a driving trip my father introduced me to the following whimsical practice. After the meal he would put a quarter in the machine, pull the handle, then slowly walk to the entrance as though totally indifferent about the outcome. He would time his walk carefully, so if he heard the machine click and go silent he would continue his walk to the door. Of course if he heard the jingle of coins or tokens being released, he would close the door he had just opened, and hurry back to collect, much to the amusement of onlookers, including me.

When I was in high school, my father provided me with a small laboratory in a room off the kitchen. It included a microscope, a Bunsen burner, flasks, test tubes, long glass tubes, and other simple equipment. The microscope was powerful enough for me to see amoebas and paramecia swimming about in water I gathered from stagnant ponds. I made slides of fly wings, the underside of fern leaves, and other things. I constructed a Heron's fountain, a marvelous invention of the Greek hydraulic scientist Heron. The fountain actually shoots a stream of water higher than its source. I bent a glass rod into a peculiar shape to form a self-starting siphon. You put one end in water and the siphon at once starts to work!

I collected different things—butterflies that I put inside glass picture frames, and leaves that I mounted in a scrapbook with the name beside each leaf of the tree they came from. I collected stamps for a short time, and mechanical puzzles. I wrote for *Hobbies* magazine the first article ever published about collecting mechanical puzzles.

I remember with pleasure the times Dad took me fossil hunting. On an outcrop near Tulsa I actually found a trilobite. He took me to Mammoth Cave in Kentucky and to Carlsbad Caverns in New Mexico. Other trips were to drilling wells. Dad was one of the last of the so-called wildcatters, independent oil producers who searched for oil independently of any of the big oil companies.

I recall vividly a trip we made to Paris, Texas, where a terrible accident had occurred. A drill became stuck in one of Dad's wells. While drillers were increasing the pull on the drill, the entire derrick, poorly constructed, collapsed, killing one of the drillers. I'll never forget the twisted steel and the blood.

Other memories are associated with the same trip. Of course I enjoyed later telling friends I had spent a week in Paris. One sunny afternoon when I was sauntering down Paris's main street, a pretty girl looking out of a store window flashed me a great smile. She could have been a waitress. I was tempted to go inside and perhaps get acquainted, but I was too shy. I walked on after smiling back, but I have never forgotten her smile.

The mayor of Paris arranged for a gathering of local businessmen at his house to meet my father. Of course if the well struck oil, it would be a great boon for the town. (The well, I should add, proved to be a dry hole.) At the gathering lots of off-color jokes were told. My dad thought them unsuitable for me to hear, so I had to sit glumly for an hour or so on steps outside the house listening to howls of laughter within.

Three stories Dad told me about this trip have remained in my memory. He said a local farm boy came every day to the well to watch the drilling. He never said a word. After many silent visits, one day he started to speak. All the drillers stopped work to listen. What he said was, "It snowed here once."

The second story he told me was about an aged farmer who lived not far away. After the derrick collapsed, one afternoon his wife said, "Something's happened at the oil well. The derrick has disappeared."

"That can't be," her husband replied. "I can see the derrick as plain as day."

I used this incident as a metaphor in my theological novel *The Flight of Peter Fromm*. There it symbolizes the fact that millions of liberal Christians around the world, both Catholic and Protestant, who attend church out of force of habit, think Christianity is still standing when in fact all its great doctrines are slowly crumbling. In England Matthew Arnold wrote his fine poem "Dover Beach" about the "melancholy long withdrawing roar" while the old doctrines glide out of Anglican churches. It is said that Archbishop Benjamin Jowett would join in reciting the Apostle's Creed by saying "I" in a loud voice, followed by "used to" in a whisper, then back in a loud voice with "believe."

My third memory concerns a time when Dad drove a black roustabout, as drillers are usually called, from the well to town. The wind blew a paper bag across the road in front of the car. My father was much amused by the man's comment. "That bag," he said, "crossed the road just like somebody."

My father was always quick to note something funny or interesting about a chance remark. As I type, there comes to mind another incident when Dad and his accountant Harry White were in the Tulsa post office to check the company's mailbox. "Was there anything in the box?" my dad asked. Harry replied, "Nothing but a notice that the box rent's due."

"Harry," said my father, "you are an unintended poet. You just spoke in perfect meters."

My father surprised me one day by reciting the alphabet backwards. I have no idea why as a boy he had bothered to memorize a backward alphabet. Perhaps it was just a challenge to his memory. He had a good memory for poems he liked, and could recite large chunks of poems by Robert Burns. Burns was one of his favorite poets, as he is of mine:

Ye flowery banks of bonnie Doone,
 How can ye bloom so fair?
How can ye chant, ye little birds,
 And I so full of care?

One afternoon when Dad and I were walking
through a meadow in Kentucky, he pointed to a small
hole in the ground. At the bottom of that hole, he
told me, lives a peculiar insect. To prove it, he found a
small stick, poked it into the hole, then quickly pulled
it out. Sure enough, to my surprise, a funny-looking
bug was clinging to the twig.

With much regret I feel I now must in honesty
report a darker side to my parents. Although Kentucky was a border state during the Civil War, both
parents believed that blacks were intellectually inferior to whites. Not that they didn't get along well with
occasional black cooks my mother hired. Mother was
proud of her Southern heritage. Whenever she was
at an occasion when a band played *Dixie*, she would
always stand. She liked to complain about the bad rap
slave owners got for the way they treated slaves. The
owners, she would say, for the most part were kind
and generous to their slaves. When Mother was in her
nineties, she did an extraordinary thing. She wrote
her own long obituary and sent it to one of Tulsa's
newspapers. They printed it on the front page without
altering a word.

For several summers, when I was small, my mother
would rent for six weeks a house in Chautauqua, a
resort town in New York. The town featured a great
variety of activities, lectures, and classes, all free once

you paid for the season or for a shorter period of time. My father liked to describe the village as filled with old ladies and their mothers. I have two vivid memories of Chautauqua. One was seeing the magician John Mulholland perform in the amphitheater. He was then editor of a magic periodical called the *Sphinx* to which I had contributed a few items. I chased him down after the show and we had a pleasant session talking about magic.

The other memory is of an evening at the town's amphitheater with my mother (Dad stayed in Tulsa for the summers to keep his oil company going), listening to a concert by the New York Symphony Orchestra. A power failure suddenly cut off lights. Unable to proceed, the orchestra leader announced in the darkness that the band would now play something they all knew by heart. It was Sousa's "Stars and Stripes Forever." When the lights came on, the trombone players were all standing on their chairs!

I could add several pages about my brother, Jim, and sister, Judith, and their bright and beautiful daughters, Dorrie on my sister's side and Cindy on my brother's side. In college Dorrie actually wrote a term paper on randomness in quantum mechanics! And there are my two great in-law nieces, Susan and Judy. I decided that to tell about my four nieces and their children would be giving readers more on relatives than they cared to know.

One summer my father took me to New Harmony, Indiana, where I walked through a reconstructed hedge maze that was supposed to resemble a maze grown by a German Adventist cult. Its twisted paths symbolize the twisted paths of sin. I marveled at the large bare footprint in stone of the angel Gabriel who the cult leader George Rapp claimed had visited him. I recall a lecture by a local historian who told how the town had earlier been a utopian colony called Harmony, founded by the Welsh socialist Robert Owen. The town had a general store called a "time store" because prices of goods increased if customers took long times to make a purchase.

I learned how Rapp had taken over Harmony after the socialist colony crumbled from bitter disputes among members. Rapp changed the town's name to New Harmony. At age eighty-nine, ill in bed, Rapp said that were he not certain God had prepared him to present his flock to Jesus at the time of the Second Coming, he might think this was his last hour. So saying, he died.

Among Rapp's eccentric beliefs was that although God was both male and female, his Son was sexless and had no sex organ! Because Rapp prohibited sexual intercourse and children, the Rappites slowly dwindled until they faded away. Byron devotes a stanza to this in his *Don Juan* poem. You'll find the stanza quoted in the footnotes of my *Whys of a Philosophical Scrivener*.

It's not easy to believe, but when the famous German theologian Paul Tillich, after several years as professor of theology at the University of Chicago, died and was cremated, his ashes were buried in a Paul Tillich Park in New Harmony! A bronze bust of Tillich is on display in the park, along with granite monuments on which are engraved quotations from Tillich's writings. You'll find photos on Google of the bust and monuments.

Paul Tillich Park was created by art patron and philanthropist Jane Blaffer Owen as a tribute to Tillich, whom she greatly admired. She is the widow of Kenneth Dale Owen, geologist and founder of the Gulf Shore Oil Company, and a great-grandson of Robert Owen. Both Robert and Jane were former residents of New Harmony. As I write (2009), Mrs. Owen, in her nineties, lives in Houston. There are entries on Google on both her and her husband.

Late in life my mother enrolled at Tulsa University to study painting under Adah Robinson, mentioned in chapter 1. Mother's watercolors, mainly of still life, were displayed in an exhibit that got favorable reviews in Tulsa's two papers. Her pictures now hang on the walls of relatives and friends. Two of my favorites that I own are a picture of many types of seeds in a vase, and an amusing painting of a dozen of Mother's lady friends around a large dinner table. Brother Jim, also late in life, took up the painting of oil landscapes.

I close this disheveled chapter with some sage advice from my father. After the wedding in Tulsa of sister Judith to James Weaver, I asked Dad if he had any good advice to give the groom. He thought for a few moments, then said, "Keep your mouth shut."

# 12

## THE NAVY, I

AFTER HITLER SNATCHED FRANCE, A SIMPLEMINDED
Stalin, like Neville Chamberlain in England, signed
a peace treaty with Hitler. All over the Chicago campus student Communists were wearing buttons that
said, "The Yanks Are *Not* Coming." When Hitler, true
to form, invaded Russia, the buttons vanished, and
the Communist Party line instantly shifted to urging
the United States to declare war on Germany.

After a draft got underway, the army classified me
4F because I was underweight. Convinced that entering the war was justified, I tried enlisting in the navy.
To my surprise I was accepted. Aware of my job in
press relations, the navy made me a yeoman, allowed
me to skip boot camp, and packed me off to the navy's
radio training school at the University of Wisconsin,
in Madison. I was put in charge of public relations and
made editor of the school's weekly paper, the *Badger
Navy News*. Each week I delivered the paper to all the
dormitory rooms and to members of ship's company.

Editing the four-page paper was a one-man job, and
very enjoyable. I wrote everything except a back page
of reports by men in training. As a member of ship's
company I had freedom to come and go as I pleased.
I bought a used bicycle. I acquired a canoe for pad-

dling on beautiful Lake Mendota. I met members of the university's faculty and visiting artist John Steuart Curry. He was then finishing a mural landscape for one of the campus buildings. The scene included a rainbow on which Curry had mistakenly reversed colors. Of course he had to repaint the bow.

I had a pleasant exchange of letters with English professor William Ellery Leonard, a famous critic and poet. Stephen Vincent Benét called his *Two Lives*, a collection of 250 sonnets, the greatest American poetic work of the century. It tells of his tragic marriage to a woman who killed herself in 1910 shortly after her wedding.

I checked out of the university's library Leonard's autobiography, *The Locomotive-God*. The title refers to one of his two curious phobias. He had a strong fear of locomotives, and he developed acute anxiety attacks if he ever left the campus or the region close to home.

One of Leonard's many books is a translation of *On the Nature of Things* by the great Roman poet Lucretius. This long poem encapsulates all the known science of the time. It includes a description of mirrors that fail to reverse left and right. I wrote to Leonard about his translation of a passage about the mirrors, which had some sentences I found obscure. Somewhere in my files I probably have his generous reply. I regret we never met. His house is now a historic spot for Madison visitors.

The head of the student union, which ran a rathskeller where students congregated for beer and snacks, was a man named Porter Butts. I like to tell people that a highlight of my stay in Madison was when I had the pleasure of introducing Butts to a sailor whose last name was Pots.

When *Life* magazine sent a photographer to Madison to do a feature on the naval training school, he brought with him the writer, who turned out to be Don Morris, my old friend from press relations days at the University of Chicago. It was a great reunion, and the last time Don and I were together.

After two years at Madison, the navy decided to transfer me to sea duty. When I boarded the USS *Pope* destroyer escort (DE-134), the sailors were buzzing with scuttlebutt about the fate of the ship's gay captain. He had been led from the ship in chains. Several young seamen in the crew had informed the navy of sexual molestation by the skipper. He had been replaced by a former druggist who proved to be an excellent "Old Man," much liked by the crew.

Officers and chiefs were also the finest of fellows. Chiefs are the men who actually run a ship. If all its officers were to vanish, the ship would get along very well with the chiefs in charge. Years later, when I read Herman Wouk's novel *The Caine Mutiny*, and saw the movie starring Humphrey Bogart as the paranoid captain, I was amused by the fact that chief petty officers are nowhere to be seen in the book or on the screen. Enlisted men are there, but only for comic relief. Of course I write from the enlisted man's point of view.

I have many happy memories of the USS *Pope* and its merry crew. I sometimes dream I am back on the ship. I can close my eyes and in my mind walk over the ship, go through its hatches, climb up and down its ladders. The ship was one of six identical destroyer escorts that together roamed the Atlantic looking for

German submarines. We had sonar equipment for finding them, and depth charges for sinking them. Before the war ended, one of our ships was sunk by a German torpedo. Many of the crew were killed by explosions of the ship's depth charges. There had been no time to set them on safe.

I was known to shipmates as Buzz. The nickname came about as follows. All my life I never, to my regret, had a nickname. The first time a sailor asked my name, I decided on the spot to give myself a nickname. For reasons I now can't recall, Buzz came to mind. It passed quickly around the crew, and for the rest of my duty on the *Pope*, both enlisted men and officers called me Buzz. A yeoman called Buzz appears in the navy chapters of my novel *The Flight of Peter Fromm*. Peter's letters to Homer Wilson, the book's narrator, provide several chapters that are fairly accurate accounts of my life at sea.

Although at the time I longed for the war to end, my life on the ship was unusually free of anxiety in spite of the fact that at any time we could be blasted by a torpedo. When I enlisted, I did not tell the navy I suffered from occasional bouts of visual migraine that could send me to bed for hours until the zigzags faded. To my amazement, I never had a single migraine headache during my four years in the navy! I attribute this to the enlisted man's freedom from having to make important decisions. You simply do what you are told to do. You do not even have to make trivial decisions, such as what shirt to put on or what tie to wear. I suspect this is one reason why so many men, in all branches of the military, decide to reenlist.

In chapter 4 I described my youthful puzzlement over the nature of my ailment. It was at Madison that I finally learned that the initial blind spot, followed by the zigzags, had nothing to do with the eyes. In the university's library I found a book titled *Nervous Diseases of the Eye*. It contained actual pictures of the zigzags. You can imagine my relief at knowing I was not going blind! I still have the visual attacks, but only once every three or four months, and now they last only about twenty minutes.

Of course I had to learn how, in speaking to shipmates, to toss in the four-letter words. We had one career old-timer aboard who managed to insert the F-word *between* syllables. For example, he would say "I guaran-f——tee it." On one occasion he approached me after learning that I liked to do card tricks, and said, "Do you know that trick when you let someone pick a card, then you tell him what the goddam f—— son-of-a-bitch is?"

I still remember an occasion when I played a joke trick, well known to magicians, on one of the sailors. I asked him to take any card from the deck and hand it to me without looking at its face. I held up the card, face toward me, and asked him to name any card. He said, "Jack of hearts." The card happened to be the jack of hearts, so I turned it slowly around and saw his face turn beet red. No doubt he has told his grandchildren how a yeoman on the *Pope* had worked a miracle.

The "out," by the way, when the card *doesn't* match the call, is to say, "Correct! Now I'm going to do something *even more* amazing. I'm going to transform the card you named to . . ."; then you name the card you are holding. Of course the trick is a whopping success only once in fifty-two times on the average, but when it fails, as it does about fifty-one times, the spectator thinks the whole thing was an intended joke.

My best friend on the ship was yeoman Vernon Pietz. After the war he landed a cushy job with Chicago's famous Museum of Science and Industry. Vern was the only shipmate I kept in touch with after the war. We shared a fondness for Dixieland music—progressive jazz buffs call us "mouldy figs." When our ship docked at the Brooklyn Navy Yard for repairs, we would enjoy the music of a Dixie band at a spot in the Village called Nick's.

The bandleader at Nick's was Mugsy Spanier, who played trumpet. Miff Mole handled the trombone, and Pee Wee Russell was the clarinetist. When I mentioned to Pee Wee that I enjoyed his solo on a recording called "Clarinet Blues," he feigned astonishment and said, "You listened to that record and you're still alive?" Later I wrote several short-short stories about jazz that appeared in an obscure little magazine. You'll find them in my short story collection *The No-Sided Professor*. Several years after the war I received a sad letter from Vern's wife telling me of his suicide. I never asked her for reasons or details.

Once on an officer's desk, at a naval base, I saw a sign that said, "Wait—there's a harder way!" It's true that the navy's red tape can infuriate a yeoman. It won't do, as an old joke goes, to substitute blue tape. However, there *are* shortcuts I soon discovered. Let me cite two.

One day I was preparing papers requesting an electrician's mate for the ship. The papers would go up the chain of command, then down the chain to some office where the request would be considered. The process could take weeks. A chief yeoman who was on board for a brief time after I joined the *Pope* saw what I was doing.

"No, no, Buzz," said. "There's a simpler way."

The chief led me to a building at the Norfolk naval base where our ship was anchored. We climbed steps to an office, where the chief greeted a sleepy yeoman by first name. "We need an electrician's mate," he said. "Is there one you can give us?"

The yeoman disappeared for a few minutes, then came back with some papers. "We have three."

"Could you send us one?"

"No problem. I'll get him to you tomorrow with the papers to sign."

My second example of tape cutting concerns the navy's free book program. During the Second World War the navy decided to print cheap paperback editions of both fiction and nonfiction that could be obtained free by anyone in the service. I had been trying vainly to obtain a book requested by one of our officers. One afternoon, when the ship was tied up at Norfolk, I walked past a building with a sign that suggested it was a warehouse for the books. I asked the sailor on duty if I could obtain a copy of a certain book.

"Sure," he said. Then he added, "We're having trouble getting rid of books. There's a wheelbarrow parked outside. If you like, you can take as many books as you please. Does your ship have a bookcase?"

"It does," I said, remembering an empty bookcase in a corner of the sleeping quarters.

I spent a half hour selecting titles. Our enlisted men paid little attention to the books, but the officers were delighted. One ensign was especially glad to get a textbook on plumbing. He had always wanted, he told me, to be able to do his own plumbing repairs.

# 13

## THE NAVY, II

ASIDE FROM BERMUDA AND CUBA'S GITMO BAY, EN-
GLAND was the only foreign nation I was able to visit
during my sea years, and then only the city of Liverpool. Many of its buildings were still in ruins from Hitler's bombs. I became familiar with the pubs on Lime
Street where the hookers congregated. I recall helping
a young prostitute put on her overcoat. She said something in a dialect that I interpreted as "Sir, your kind-
ness is crushing." At a great used bookstore I bought
several books by William James, which I mailed to my
parents to keep for me. The young woman who sold
me the books had a great smile that revealed several
missing front teeth. Believe it or not, I had learned
somehow that Liverpool was headquarters for England's Auguste Comte Society. Thinking its building
might have a museum or a shop, I actually walked to
the building, only to find it locked.

The *Pope* had an unintended clown on board whose
name I withhold in case he had children who might
be living. Call him X. He was constantly asking for
a transfer to land on the grounds of frequent bed-
wetting. One night, when as usual we were sailing
darkened ship (even a match flame on deck might be
seen by a German sub), X was on duty as a bridge

lookout. Suddenly he turned on a huge searchlight! He wanted to see the time on his wristwatch!

We had a mascot on board, a friendly mongrel dog named Seaweed. One day, after being kept awake by Seaweed barking, I said to X that the dog was becoming a nuisance, and someone should toss him overboard. Next day Seaweed vanished. To this day I have twinges of conscience because I suspect X had taken me seriously. X's father was a southern farmer. One day X showed me a letter from his dad. It was signed, "Sincerely yours," followed by a full name signature.

In some ways X was far from stupid. Occasionally a motion picture would be shown on the fantail. Of course it would draw a large audience. To obtain a close-up seat, X found he could sit behind the screen and see the film in mirror image form!

When not working in the yeoman's shack, I had two other duties. One was steering the ship during the night midwatch, and the other was sitting on the bridge as a starboard lookout. I have happy memories of white foam rushing past the ship's sides, the wind in my face, and that great line in mind from John Masefield's "Sea Fever," " . . . the flung spray and the blown spume, and the sea-gulls crying." There always were seagulls flying close to the ship, waiting for garbage to be tossed overboard.

DEs, I should mention, are small enough to be constantly swaying in choppy seas. After going aboard, I was so sick for three days that I considered going AWOL (absent without leave) as soon as the ship docked, but after several days I felt great and was never sick again. Indeed, the rougher the sea, the more I liked it because there was less danger of being hit by a torpedo.

Before I joined the *Pope*, the ship had actually captured a German sub intact and towed it to the United States! It was the first time such an event had occurred. Of course it was kept top secret. The captured sub is now on display at Chicago's Museum of Science and Industry.

Another time, a German submarine, severely crippled by our depth charges, surfaced. Rather than have his sub captured and towed to the States, the captain ordered abandon ship, then sank his submarine. Its crew was distributed among our set of sister ships. Later, after the German surrender, a submarine surrendered to us, and we escorted the ship and part of its crew to our home base at Norfolk.

I remember how amazed our crew members were to discover that the German sailors were just like us, good-natured youngsters, happy the war had ended and anxious to make friends with our sailors. The German officers were a different breed. They were astonished that some of our officers were Jewish, and that there were even black sailors on board. It was not until after the war that the navy became fully integrated. Throughout the war blacks slept in a private corner of the ship and were servants to the officers.

Our ship's chief radioman, Sam Hall—he was in charge of our secret code machine—turned out after the war to be head of Alabama's Communist Party! Sam and I became good friends. When I was assigned during the midwatch (12:00 to 4:00) to steer the ship,

Sam would join me in the pilot room to chat about politics. I knew he had left-wing views and admired Russia, but I did not know he had been an active CP member until I saw a story about him, with his picture, in the *New York Times*. I was told he died a few years after the war.

When news came over the intercom about the dropping of two atom bombs on Japanese cities, I was the only person on board who realized the war with Japan had ended, and that the world had entered an atomic age. I knew about our work on the bomb because anyone in press relations at the University of Chicago knows everything happening on campus. I knew that Enrico Fermi and his men were working on the bomb in a laboratory under Stagg Field. In the morning, when I left my sleeping room to walk to the campus, Fermi would pedal past me on his bicycle.

A Tulsa friend, Betty Murray (a chapter on Bob and Betty Murray will come later), worked during the war as a secretary near the campus for what was known as the Manhattan Project. One day a lady at a desk close to Betty's said, "I wish I knew what this project is all about."

"Don't you know?" Betty replied. "They're trying to build an atom bomb."

A hush fell over the room. Next day Betty was visited by an FBI agent who wanted to know how come she knew that. Betty said, "Why, everybody at the university knows that." She may even have mentioned me. At any rate, no one from the FBI came to see me.

When our two bombs ended the war with Japan, the *Pope* was not shifted from the Atlantic, where it always operated, to the Pacific. In fact, the ship was no longer of any value to the navy. Along with its four sister ships that survived the war, the *Pope* was sent to Green Cove Springs, Florida, for decommissioning.

Vernon Pietz and I were then the two yeomen aboard. At the time a shabby little carnival was playing on grounds near the naval base. I remember how amused Vern was when I explained the "gaff" on a carny game that involved swinging a heavy ball at the end of a chain so the ball missed a pin but knocked it over on the swing back. The game was what carnies call a "two-way store," meaning the operator can always let a mark win on a practice try, but the mark is sure to lose after paying for a turn.

At Green Cove, Vern was either transferred elsewhere or discharged, I forget which, leaving me in charge of decommissioning paperwork. It was then that I wrote a long poem called "So Long Old Girl" about the history of the *Pope*. You can find the poem in my book *The Jinn from Hyperspace*.

After my discharge I took a bus to Tulsa with a stopover in New Orleans to hear some Dixie. To find out what bands were playing where, I visited an office that handled bookings for jazz bands. It was there I met the legendary Bunk Johnson, back from New York City where he had played trumpet with a group of elderly black friends. Bunk had only recently been discovered working in the rice fields of New Orleans and in dire need of false teeth. I had a great chat with him. To prove how agile he was in his old age, he sat on a chair and put both legs behind his neck!

A girl in charge of the office asked me what instrument I played. I half lied by replying "the trombone." Actually I had found on the ship a battered trombone someone had discarded, and in one of the ship's vacant lower rooms I taught myself how to play simple tunes. Today, when there is nothing better to do, I amuse myself by practicing on my musical saw!

I spent two nights at a cheap hotel in the French Quarter. I recall stopping to chat with a young man who was standing in front of a cabaret doing his best to persuade persons walking by to come inside and see the floor show. I mentioned that I thought every taxi driver in the city was a pimp because so many stopped to ask if I wanted to see a hooker. "It's true," he said. Then he smiled and added, "I'm a pimp myself. World you like to meet one of the chorus girls?"

Soon I was back in Tulsa, reunited with my parents and able at last to remove my blue sailor suit. I shaved off a huge black mustache I had grown before I left the ship.

# PHOTO ESSAY

Martin, age three.

Martin, age five.

Martin, age six.

Martin, age six, and his
brother, Jim, age three.

2187 South Owasso, Tulsa, OK, ca. 1922.

Martin, age twelve; sister, Judy, age two; and Jim, age nine.

The Gardner family ca. 1927. Judy, age three; Willie Spires Gardner; Martin, age thirteen; Jim Jr., age ten; and James Henry Gardner.

Martin demonstrating a handkerchief character to his sister Judy.

The Gardner family, ca. 1952. Top row, left to right: Martin Gardner, James Henry Gardner, and James Gardner, Jr. Bottom row: Cindy Gardner and her mother, Marjorie Anderson Gardner, Judy Gardner Weaver and her son, Teddy Weaver, and Willie Spires Gardner.

Martin and Charlotte Greenwald, Central Park,
New York City, ca. 1950.

Mr. and Mrs. Gardner, New York City, wedding photograph.

Jim, Charlotte, Tom, and Martin Gardner, Dobbs Ferry, NY, ca. 1961.

Dr. Martin Gardner (Honorary Doctor of Humane Letters) and Charlotte Gardner, Bucknell University, May 1978.

Martin and Charlotte in their living room at 10 Euclid Avenue, Hastings-on-Hudson, NY, 1980.

Martin's office consisted of the entire third floor of 10 Euclid Ave., Hastings-on-Hudson, NY.

Publicity photos of Martin posing with *Alice in Wonderland* characters statue, Central Park, NY, 1960, for the first edition of *The Annotated Alice in Wonderland*. Photos by George Cserna.

Publicity photo of Martin posing with the Alice statue thirty years later, 1990, for his sequel, *More Annotated Alice: Alice's Adventures in Wonderland & Through the Looking Glass*. Photo by Scot Morris.

Caricature of Ronald Crane
by Martin Gardner, *Phoenix*
magazine, February 1937.

Caricature of Charles Morris
by Martin Gardner, *Phoenix*
magazine, March 1937.

Terrible Mike illustration by Martin Gardner. Appeared in "The Ballad of Terrible Mike," *Phoenix* magazine, January 1937.

"The Maid," by Martin Gardner. Appeared in "The Ballad of Terrible Mike." *Phoenix* magazine, January 1937.

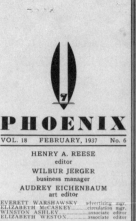

# PHOENIX

VOL. 18    FEBRUARY, 1937    No. 6

**HENRY A. REESE**
editor

**WILBUR JERGER**
business manager

**AUDREY EICHENBAUM**
art editor

EVERETT WARSHAWSKY....advertising mgr.
ELIZABETH McCASKEY....circulation mgr.
WINSTON ASHLEY....associate editor
ELIZABETH WESTON....associate editor

### EDITORIAL ASSOCIATES

V. P. Quinn             Theodora Schmitt
Margery Goodkind        William Crockett
C. Sharpless Hickman     Jean Garrigus
Dick Lindheim           Harvey Karlen
Meyer Becker            Martin Gardner
David Eisendrath        Sam Hair

### BUSINESS ASSOCIATES

Harker Stanton          Eleanor Cupler
Harry M. Hess           Mary Ann Patrick
Dan Heindel             Betty Quinn
Beryl Lazar             Mary Letty Green
Robert Warfield         Bud Daniels
Mel Rosenfeld           Franklin Horwich

The University of Chicago official student magazine, established 1919.   Published monthly (except July and August) by and for the students
of the University of Chicago.   The University
is not responsible for any material herein, nor for
any contractual obligations.   Fifteen cents the
copy, one dollar the year from September to June.
Entered as second class matter, November 12, 1936,
at the post office in Chicago, Illinois, under the
Act of March 3, 1879.   The contents are not to
be printed without permission.   Mailing address,
Faculty Exchange, Box Ninety-seven University
of Chicago. Telephone Dorchester 7279.

## CONTENTS

Last year Martin Gardner edited Comment, gentle literary mother of this year's Phoenix. This year he contributes to the erratic son. He seems to be a caricaturist, magician, writer, and philosopher, though he does everything so quietly that you can never be sure—until he crashes through with something like the Crane Caricature on page 21. Then you know he's a caricaturist. Sit around until your watch disappears and you know he's a magician. Wait until he gets a story in to Phoenix which he's been promising for three months and you'll know he's a writer. Check up on his registration and let him talk privately to you of the Aristotelians, and you know he's a philosopher.

He has an intense interest in his surroundings and likes to wander through old book-stores and North Clark Street. Hailing from Tulsa, Oklahoma, he has somehow managed to acquire a wide and varied acquaintanceship in Chicago. If you want to know of an obscure but excellent place to eat, drink, or be otherwise entertained, Gardner will know just the place.

He's a small fellow with high narrow shoulders, an elastic mouth, and deep-set, dark eyes. That's he up above.

Caricature and futuristically insightful blurb introducing Martin to the readers of *Phoenix* magazine, February 1937. Artist unknown.

10 EUCLID AVENUE
HASTINGS-ON-HUDSON
NEW YORK 10706

Martin Gardner regrets that it is impossible for him to:

1. Evaluate:

   Angle trisections

   Circle squarings

   Proofs of Fermat's last theorem

   Proofs of the four-color theorem

   Roulette systems

2. Give advice on, or supply references for, high school science or math projects.

3. Inscribe books for strangers.

4. Give lectures, or appear on radio or TV shows.

5. Attend cocktail parties.

6. Make trips to Manhattan except under extreme provocation.

7. Donate books to libraries.

8. Provide answers to old puzzles.

9. Prepare material on speculation for toy companies or advertising agencies.

10. Put the reader in touch with Dr. Matrix.

Form letter used by Martin to reply to readers' requests.

The world's most skeptical man! Cartoon by Joe Nickell for the 1996 Gathering for Gardner.

One day while out walking, Peter met Humpty-Dumpty. He was sitting on his favorite wall, up which he tacked a puzzle sum for Peter to solve. He said it was the name of a dear friend of his, a girl you know too. Who is the girl? And can you find the mistake in this picture? Source: *Peter Puzzlemaker* (Dale Seymour Publications, 1992), p. 28. (Solution: Cone + Stall = Conestall + Ink = Conestallink − link = Conestal − Nest = Coal − Co = Al + Ice + Alice. Mistake: Humpty-Dumpty has two left ears.)

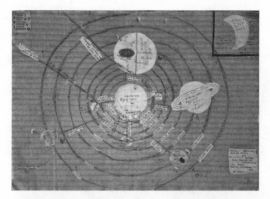

The Solar System,
a science project
by Martin, age ten.

Oblivious to a stripper, Martin (far left) concentrates on a magic trick
as part of a publicity photo for the 1947 Society of American Magicians
conference. Photo by George Karger/TIME & LIFE Images/Getty Images.

# Portraits of Martin

Numerous 3-D illusions in the foreground, Martin, and David
Eisendrath's photo of Einstein in the background. Photo (of Martin)
© Elliott Erwitt/Magnum Photos.

From *Magic* magazine.

A version of a classic illusion in tribute to Martin. Artwork by Victoria Skye.

Martin in person and in dominoes. Domino portrait by Ken Knowlton.

# Martin and friends

Martin and a friend, ca. 1950.

Jay Marshall and Martin Gardner, ca. 1960.

Bruce Elliot, Martin Gardner, Clayton Rawson, and Jay Marshall, at a magic convention, June 1960.

Martin signing copies of his *Annotated Alice in Wonderland* books at the opening of the Woodland Hills, CA, Fry's Electronics. This store's interior decoration theme is Alice in Wonderland; this is one of the few times he participated in a public book signing.

Martin and Ray Hyman, 1996.

Martin and friends at the Gathering for Gardner: John Conway, Solomon Golomb, Robert Darling, Martin, Dick Hess, and Nob Yoshigahara.

James Randi and Martin, ca. 2000.

Bob Murray.

Tom Rodgers (originator and host of the Gathering for Gardner) and Martin.
Norman, OK, 2009. Photographer unknown.

Martin standing in front of a bookcase filled exclusively with copies of books he wrote. The bottom two shelves, which are not visible, are also full. March 2006. Photo by Colm Mulcahy.

# 14

## ESQUIRE AND HUMPTY

AFTER THE WAR AND A BRIEF STAY IN TULSA, I MOVED back to the University of Chicago area to a dreary single room off Fifty-Fifth Street. My window opened on an air shaft. When my sister, Judith, visited me, she was so shocked by how dirty the window was that she insisted on giving it a scrubbing. My only possessions then were an alarm clock, some books, but no radio. I kept notes of my reading and speculations on three-by-five file cards that I kept in ladies' shoe boxes, which I picked up free from shoe stores. When I cut and pasted down paragraphs from books, they went on four-by-six cards that I put in men's shoe boxes. I kept the boxes in the room's closet.

Many years later when my old friend John Shaw, the Sherlock collector, visited me, he looked through my four-by-six cards to see what I'd been reading. He was shocked to discover I had chopped up and pasted down passages from a rare first edition of *The Great Gatsby*! Of course it was not so rare when I bought it.

I could have gotten back my old job in press relations had it not been for an event that was a huge turning point in my life. I actually sold a short story to *Esquire*! Called "The Horse on the Escalator," it was about a man who collected shaggy-dog jokes about horses. Sample:

A man tries to get on a Marshall Field's elevator with a horse.

"Sorry," says the elevator girl, "you can't bring that horse on the elevator."

"But lady," the man replies. "He gets sick on the escalator."

After the story was published, I sent *Esquire* a letter signed with a fake name and address. *Esquire* printed it. It said how much I enjoyed the story, and added a horse joke my tale had missed.

*Esquire* got a batch of authentic favorable mail about my story. Someone told me that Red Skelton mentioned it on his radio show. As a result of this reader reaction, *Esquire*'s editor Fred Birmingham invited me to lunch. It was at a fancy Chicago restaurant. I recall how the hat-check girl winced and almost held her nose when she took my old navy pea jacket. It smelled strongly of diesel oil, the result of a time when new oil overflowed the *Pope*'s tank and streamed into our lockers.

Fred asked for a second story. I obliged with my best-known science-fiction yarn, "The No-Sided Professor." I had become fascinated by topology, a branch of mathematics which studies the properties of objects that remain unaltered no matter how the object is twisted or distorted. It is sometimes called rubber-sheet geometry because regardless of how you stretch or twist a rubber sheet on which figures are drawn, the figures' topological properties will not change. For example, a closed curve will always divide the sheet into two parts, inside and outside. Called the Jordan theorem, it is obvious but not so easy to prove. And there are other more subtle topological properties of the curve.

My second *Esquire* story involved a topologist I called Professor Slapenarksi. It concerned his discovery of a variation on the Möbius strip. An ordinary strip has two sides and two edges, but if you give it a half twist and join the ends, it becomes one-sided and one-edged. Slapenarski discovered a way to make a *no-sided* surface. When you joined the ends of the strip, the thing vanished! My story got wilder when Slapenarksi had a fight with his rival, a mathematician named Simpson. He folded Simpson into a three-dimensional model of his one-sided surface, and poor Simpson disappeared, leaving only a heap of clothes. A French magazine translated my crazy tale, and it found its way into several science-fiction anthologies.

Fred asked for more stories, and for a year or two I lived on income from *Esquire*. With few exceptions all my *Esquire* stories are in my collection *The No-Sided Professor, and Other Tales of Fantasy, Humor, Mystery, and Philosophy*.

*Esquire* suddenly changed ownership, moved its offices to Manhattan, and Fred was no longer editor. The new editor did not care for my strange brand of humor, so my market vanished like Professor Simpson. However, now aware that I actually could get paid for writing, I decided to move to New York City, where most periodicals are edited and so many books are bought and published.

Although I made a few paltry sales to little magazines, it was impossible to live on this income. Of course I collected dozens of rejection letters and postcards. The worst story that came back was called

"Occam's Razor." Long ago I threw away the manuscript, and although my memory now is hazy, the plot went something like this. William Occam was a brilliant but eccentric young man who lived on Manhattan's West Side not far from Central Park, supported by a wealthy father in Omaha. Bill was obsessed with a desire to simplify his life. He lived in a small bedroom with as few possessions as possible. His shabby clothes and sandals came from a Salvation Army store. An admirer of Thorstein Veblen, Bill owned no watch, no ring, no radio, only an old Underwood typewriter. Like Thoreau, another of his heroes, he kept careful records of the life—animals, birds, even insects—that he observed daily in Central Park, He was writing a book about the park that he called *Manhattan Walden*.

Life began to get complicated when he fell in love with an artist in the Village. Soon they were married. Forced now to get a job, he could find nothing better than a position as an errand boy at Random House, a job he despised. He became more and more depressed. The final blow was when his wife gave birth to triplets. Bill ended his miserable life by slashing his throat with a razor blade. My final sentence was something like "The cut was a perfect geodesic, the shortest and simplest curve joining two points on his neck."

I did get a nice rejection note from William Barrett, then editor at *Partisan Review*. He liked my paragraphs about Central Park but found the rest of the story worthless. Of course he was right. My only other written rejection was from Gershon Legman, then editor of *Neurotica*. He said the magazine printed no fiction. Later he and I became friends, as I tell elsewhere.

Luckily a pal came to my rescue. He was Harold Schwartz, recently hired by Parents' Institute to start a line of pulp magazines for youngsters. One of the new magazines was *Humpty Dumpty*. It was to be edited by none other than Humpty himself, who had a small son called Humpty Dumpty Junior. The magazine's name had been proposed by the wife of the owner of Parents' Institute. She also suggested that each issue contain a story about the adventures of Junior, and a poem of moral advice from Humpty senior to his son. It was assumed that both story and poem would be read aloud by a parent.

Harold, bless him, hired me for a job as contributing editor. For eight happy years, most of the time working at home, I wrote Junior and the poem, and also provided each of the year's ten issues (summer months were skipped) with the magazine's activity features of the sort that destroyed pages. You folded a page to change a picture, or held it up to a light to see something on the back of a page, or pushed something through a page (e.g., the blades of scissors to make a crane open and shut its beak), or moved a strip back and forth through slots, or rotated a circle pinned to a page, and so on.

For example, in the first Christmas issue I had a feature that involved pushing Santa down a chimney; then you turned the page and extracted him from the fireplace. A later feature showed Tony eating spaghetti. You put a piece of white string behind the page, then poked one end of the string through a hole in Tony's mouth. By pulling the string into Tony's mouth you could see him slurp up the spaghetti.

At that time Shari Lewis had a popular television show on which she did superb ventriloquist work with a hand puppet called Lamb Chop. Shari also liked to introduce amusing stunts and simple magic tricks. (Her father was a professional conjuror who worked in New York City under the name of Peter Pan.) Shari was so much amused by my Tony feature that she had it enlarged and demonstrated on her show!

My inspiration for activity features was George Carlson, the art editor of a children's magazine called *John Martin's Book*. During the twenties it was the delight of my childhood, mainly because of a Carlson page called Peter Puzzlemaker. On each page Peter introduced a simple puzzle illustrated by a scene in which an amusing mistake was made. I remember how eagerly I worked on each puzzle and searched for the mistake before answers appeared in the next issue. The mistakes were cleverly concealed, but obvious once you spotted them, such as a candle with tallow dripping upward instead of down, or a star within the horns of a crescent moon. Carlson also did features that involved cutting or folding a page. You can say that for *Humpty* I took up where Carlson left off.

Carlson was an interesting artist who deserves to be recognized. He did lots of covers for *John Martin's* magazine. He wrote and illustrated many books for children. He did the front cover of the jacket for the first edition of *Gone with the Wind*. If you care to know more about Carlson and his friend John Martin, see chapter 9, "*John Martin's Book*: A Forgotten Children's Magazine," in my collection *From the Wandering Jew to William F. Buckley, Jr.* Years later I edited Carlson's book *Peter Puzzlemaker* for a new edition, followed by

*Peter Puzzlemaker Returns*, a selection of pages from later issues of *John Martin's Book*.

I was never able to interest a publisher in books based on my hundreds of activity features. Publishers are understandably reluctant to publish books that no libraries will buy if their pages are likely to get demolished. I did, however, persuade Simon and Schuster to do a selection from the eighty poems I wrote for *Humpty*. The book, now long out of print, was titled *Never Make Fun of a Turtle, My Son*. The title poem goes as follows:

NEVER MAKE FUN

Never make fun of a turtle, my son,
    For moving so slow in a race.
He *prefers* to move slow and he thinks that *you* go
    At a terrible, nerve-wracking pace.

Don't ever sneer at a beaver, my dear,
    Because of the size of his tooth.
He wonders why all of your teeth are so small,
    And thinks that *your* grin is uncouth.

It just isn't fair to guffaw at a bear
    Who is brown from his snout to his feet.
He *likes* to be brown. He thinks you're a clown
    With a face painted white as a sheet.

It's vulgar to laugh at a baby giraffe.
    His neck is unusual, that's true.
But I tell you he's *glad* to resemble his dad,
    And would hate
    To be shaped
Like you!

While living in Manhattan, in my *Humpty Dumpty* days, I frequently contributed, without pay, book reviews and articles to the *New Leader*. This was a weekly democratic socialist periodical that took a vigorous anti-Communist stand at a time when it was not fashionable for left-wing journals to do so. I would visit the *New Leader*'s dingy offices, when Sol Levitas, the magazine's founder, was editor, and select a book I would like to review. One of my articles was "H. G. Wells, Premature Anti-Communist." Another, "Mr. Smith Goes to Tulsa," was on how Tulsa's "silent treatment" was handling the settling in Tulsa in 1947 of the notorious Reverend Gerald L. K. Smith. Both articles can be found in my book *Order and Surprise*. The *New Leader* lingered on until 2008, when it became a bimonthly online publication.

Harold left Parents' to start Greenwood Press, now the publisher of a large variety of scholarly journals. It began with two quarterlies I suggested to Harold, the *Journal of Recreational Mathematics*, and *Word Ways*, devoted to recreational linguistics. Both periodicals are still being published. Harold hit the jackpot by reprinting complete runs of old left-wing periodicals. These sets, with introductions by experts, were snapped up by American libraries. I was unable to convince Harold that runs of old pulp magazines devoted to detective fiction and to science fiction might do equally well.

Many of my *Humpty* poems have been reprinted in anthologies of verse for children, and one of the stories appeared on a phonograph record. Klutz Books years later paid me for a batch of activity features from *Humpty*, which they used in a book of things for a child to do while traveling. Of course illustrations for the activities had to be redone because I don't have art reprint rights.

On two occasions I worked in the offices of Parents' Institute to edit the first few issues of two other children's periodicals. One was called *Piggity's*. It was first intended to be *Piggly Wiggly*, but the same-named grocery chain stores had that name trademarked. I wrote a story in each issue about a little pig. The magazine lasted only a few issues.

The other magazine I launched was *Polly Pigtails*. I was Polly, who wrote a letter for the front of each issue to the young girl readers. I also provided unsigned filler material about things to make and pranks to play on friends. In 1941 the magazine's name was changed to *Calling All Girls*. It changed again to *Young Miss* in 1955, and to *YM* in 1986. Under that name it is still doing well.

Parents' also published, under Harold's supervision, a monthly called *Children's Digest*. I contributed to it both articles and filler material, such as puzzles and brainteasers. It, too, like *Humpty Dumpty* is still flourishing.

Like Chicago, living in Manhattan was another gratifying experience. I have a strong memory of sitting on a side seat of a subway car, Charlotte on one side and Jimmy, then about six months old, sound asleep with his head on my shoulder. People walking by invariably gave us a smile. I thought, "Here I am, in one of the world's greatest cities, with a woman I love next to me, a son sleeping in my arms. What more could I desire?" It was the happiest moment of my life.

# 15

## SCIENTIFIC AMERICAN

THE SECOND LUCKIEST EVENT IN MY LIFE—THE FIRST was meeting Charlotte—was my association with *Scientific American*. Here's how it came about.

One afternoon I was visiting a New York City stockbroker named Royal V. Heath. We were friends through a mutual interest in magic. Heath had written a little book on number tricks, and I had published a series of articles on mathematical magic in *Scripta Mathematica*. This was a journal edited by Jekuthiel Ginsburg, of Yeshiva University, who also sponsored gatherings at the university to hear mathematicians give talks on recreational topics. My articles were later made into a book, *Mathematics, Magic, and Mystery*, still in print as a Dover paperback.

Heath showed me a mathematical toy I had never seen before. It was a large cloth structure called a hexahexaflexagon. You "flexed" it a certain way that unfolded it, then folded it back again to display a face of a different color. Heath told me it had been invented and studied by a group of graduate students at Princeton University. One *hexa* in its name is for the number of sides, the other for the number of different faces that can be exposed by flexing. Heath gave me the name of one of the students, John Tukey, who later became a renowned mathematician.

I had previously sold to *Scientific American* an article on logic machines. It occurred to me that the magazine might spring for an article on flexagons. I drove to Princeton, where I met Tukey, and Bryant Tuckerman, responsible for the "Tuckerman Traverse," a way of exploring all the faces of a flexagon. Arthur Stone, the actual discoverer of flexagons, was not there because he lived in England, nor was Richard Feynman, already a famous physicist at Caltech. He had been a major contributor to flexagon theory when he was at Princeton.

*Scientific American* snapped up my article on flexagons and ran it in the December 1956 issue. All over New York City readers of the magazine, especially those in advertising offices, were making and flexing flexagons. Today there are some fifty websites devoted to flexagon theory and variants of the original forms.

Gerard Piel, publisher of *Scientific American*, called me to his office to ask if there was enough similar material to make a monthly column. I said I was sure there was. I hurried at once to the used bookstore section of Manhattan, then near the Village, to buy all the books I could find on recreational math, notably W. W. Rouse Ball's classic *Mathematical Recreations and Essays*, and several lesser works by others. I submitted my first column, on a strange type of magic square that forces someone to choose a number even though the choice seems random. *Scientific American* called the column Mathematical Games, which by coincidence had the same initial letters as my full name. The rest is history.

Writing the column for more than twenty-five years was one of the greatest joys of my life. If you look over all my columns (they are collected in fifteen Cambridge University Press books), you'll find that they steadily become more sophisticated mathematically. That was because I was learning math. I had not taken a single math course in college, although I loved the low-level math I learned in high school. And I had always been fond of recreational math ever since I was introduced to it as a boy by Sam Loyd's mammoth *Cyclopedia* of puzzles.

One of the pleasures in writing the column was that it introduced me to so many top mathematicians, which of course I was not. Their contributions to my column were far superior to anything I could write, and were a major reason for the column's growing popularity. The secret of its success was a direct result of my ignorance. Even today my knowledge of math extends only through calculus, and even calculus I only dimly comprehend. As a result, I had to struggle to understand what I wrote, and this helped me write in ways that others could understand.

One of the first eminent mathematicians to contribute to my column was Solomon Golomb. I had encountered the paper he wrote as a youth on polyominoes— pieces formed by fitting unit squares together along their edges. Sol had named them polyominoes, as well as naming subsets of using $n$ squares. A single square is the monomino, two squares are the dominoes, three the trominoes, four the tetrominoes, and five the pentominoes. The problem of finding a formula for the number of $n$-ominoes, given $n$, is still a deep unsolved combinatorial problem.

My first column on Golomb's twelve pentominoes was an instant hit. I returned to polyominoes in several later columns. Today they are a flourishing branch of recreational math. Two excellent books on the topic, one by Golomb, have been published.

Polyominoes of course have their cousins in higher spaces. In 3-space they are called polycubes, unit cubes joined by their faces. The most famous puzzle involving polycubes was invented by Piet Hein, best known in Denmark as a poet. Books containing his short epigrammatic verses, called Grooks, all cleverly illustrated by Piet Hein, have sold almost as well here as in Denmark.

Piet Hein's polycube puzzle, called Soma, consists of the seven nonconvex pieces that can be formed with three or four unit cubes. Like the seven tans of tangrams, which form a square, the seven Soma pieces will form a cube, as well as an endless variety of shapes (again like tangrams) that resemble such things as buildings, furniture, even animals. Like my columns on polyominoes, my columns on Soma and other polycube puzzles opened another vast new field of mathematical play.

The famous British mathematician John Horton Conway, now at Princeton University, was the first to prove with a friend (by hand, not by computer!) that there are exactly 240 ways, not counting rotations and reflections, to make a cube with the seven Soma pieces. Parker was the first toy company in the United States to market Soma. For a short time Parker also published a periodical devoted to Soma problems.

Several of my later columns described other Piet Hein inventions, notably his two-person game Hex, played on a board that is a pattern of hexagons. The game was independently invented here by Nobel Prize–winner John Nash, about whom the book and movie *A Beautiful Mind* were made. Nash was the first to show, by a clever argument, that the first player can always win a Hex game if she makes her best moves, although the proof tells you nothing about *how* to win. There is now a large literature on Hex and its variations, and the few known strategies for first-player wins on small boards.

Piet Hein's superellipse was the topic of another popular column. It concerns a closed curve that is midway between a rectangle and an ellipse. My column led to the construction and sale in Denmark of tables in the shape of superellipses, and an object called a superegg. The egg has the property, unlike chicken eggs, of balancing on one end. Brass versions of supereggs are still on sale in shops that carry science toys. The self-proclaimed psychic Uri Geller once issued a press release in England saying that John Lennon had given him a mysterious object he said had been handed to him by aliens visiting the earth in a UFO! Inspection of a photograph revealed that Uri was holding a superegg.

Piet Hein visited me twice, and we became good friends. On his second visit he brought along his beautiful wife. A native of Iceland, she had become one of Denmark's famous actresses. When she starred in Tennessee Williams's *Cat on a Hot Tin Roof*, she wanted to check the play's English script. It had been published here as a book, so I sent her a copy. Piet Hein's letter of thanks had a postscript saying he was enclosing a picture of his home. It took me a while to realize he was playfully referring to the back of a ten-dollar U.S. bill that bore a picture of the Treasury Building! The bill was reimbursing me for the book I had sent to his wife.

Raymond Smullyan is a top mathematician and logician who contributed to my column, and who became a good friend. I have an anecdote about him. After publishing a now-classic work on formal logic systems, he put together a collection of original chess problems, each embedded in a vignette about conversations between Sherlock Holmes and Dr. Watson. To help Ray find a publisher, I phoned my editor at Knopf. I described to her Ray's manuscript and asked if she would like to see it.

Her instant answer was "No. It's not the type of book Knopf would ever consider."

Ray decided he needed an agent. The first house the agent tried was Knopf. After seeing the manuscript, Knopf signed a contract for it.

On the phone later with my Knopf editor—call her Betty—I said, "By the way, my friend Smullyan found a publisher for his *Chess Mysteries of Sherlock Holmes*."

"Yes?" said Betty. "Who's the publisher?"

"Knopf," I said.

There was a long silence while I imagined Betty getting up from the floor.

The book did so well that Knopf followed it with *The Chess Mysteries of the Arabian Knights* and several other books by Smullyan. He quickly became a popular author. One of his books is titled *What Is the Name of This Book?* The question, of course, becomes the title, an example of paradoxical self-reference that Ray is fond of devising.

A recent Smullyan book is titled *Who Knows?* It consists of three parts. Part 1 is a lengthy, sympathetic commentary on my confessional *The Whys of a Philosophical Scrivener*. Part 2 is an attack on the Christian doctrine of hell. Part 3 is what Ray calls "Cosmic Consciousness."

Smullyan enjoys telling philosophical jokes of his own creation, always laughing heartily when he comes to the punch line. One of his jokes I especially like concerns all the great philosophers of the past, who appeared to him in a dream. Each gave a short statement expressing the essence of his philosophy. After each finished speaking, Ray said something that refuted their philosophy so thoroughly that each bowed his head and faded away in embarrassment.

Fearful that he would forget what he had said, Ray wrote down his crushing remark on a sheet of paper by the bed, then went back to sleep. Next morning he recalled his dream but could not remember his remark. He found the sheet and read these words: "That's what *you* say!"

Roger Penrose, of Oxford University, is another mathematical genius I got to know personally. I had the honor of writing the foreword to his *Emperor's New Mind*, and later reviewing his massive (1,094 pages!) *Complete Guide to the Laws of the Universe*. On one of his visits to the United States he stayed at our house in Hastings-on-Hudson. Before retiring for the night, I handed him a wooden puzzle someone had sent me that was similar to the ancient Chinese rings. The solution required hundreds of moves. In the morning Penrose handed me the puzzle solved. He had spent an hour or so on the thing before falling asleep.

I once made a trip to Winston Salem, North Carolina, to hear Penrose talk at a mathematical conference. Ed Witten, the famous superstringer, was also on the program. I understood every word of Penrose's lecture, but not a single sentence of Witten's. He kept mentioning "loop groups." I had never heard of loop groups. I asked a mathematician sitting next to me where I could find out about them. He shook his head and said that he too was unfamiliar with the term.

Roger and I share many opinions. We are both unashamed Platonists who believe mathematical theorems and objects are discovered, not created, with a reality independent of human cultures. We also agree that no computer of the kind we know how to build— that is, one made of wires and switches—will ever reach the creative intelligence of humans. And we are fellow "mysterians," convinced that at present neuroscientists haven't the feeblest notion of how our brains manage to become aware of their existences. Philosopher Daniel Dennett has written a book titled *Consciousness Explained.* He comes nowhere near explaining it, nor does his good friend Douglas Hofstadter in his book *I Am a Strange Loop.* My review, "Do Loops Explain Consciousness?" can be found in my anthology *The Jinn from Hyperspace.*

I consider Doug another friend, having given his *Gödel, Escher, Bach* a rave review in my *Scientific American* column. He is as brilliant a thinker and writer as Dennett, and although there is no question that our brains swarm with self-reference loops, the loops merely describe the way our brain operates. How they produce self-awareness remains a dark mystery. Neuroscientists are making wonderful discoveries about the brain, but it is my opinion, as well as Penrose's, that they are still far from understanding the mind of a mouse. It is no insult to neuroscience to say it is a science in its infancy.

My *Scientific American* column that had the greatest impact on mathematicians was a column that introduced the world to John Conway's famous cellular automaton game he named Life. It came about this way. On a visit to my home Conway asked if I had a Go board. I did. On the board he placed some Go stones, then explained the few simple rules by which a pattern of stones can be changed to a different pattern. The rules force some stones to "die" and be removed, and other cells to give birth to new patterns. They are called the game's "transition rules." When Life is played on a computer screen, the patterns rapidly change, sometimes leaving a blank field, at other times becoming stable or changing to "blinkers" that oscillate between two or more patterns.

Conway offered a prize of one hundred dollars to anyone who could find a starting pattern that continually added stones to the field. The prize was won by Bill Gosper, then a student at MIT. Bill had found what came to be called a "glider gun" that shoots out a steady stream of gliders. These are small shapes that glide across the page.

Conway called his game Life because it demonstrates how from a few simple "laws" complex shapes emerge that can live, move, and die like primitive life-forms. Gliders, for example, crawl across the computer screen like insects. Conway was the first to prove that Gosper's glider gun turned Life into a Turing machine that in principle can do everything the most powerful computers can do, only of course with great slowness. It can, for example, calculate the digits of the square root of 2, pi, *e*, or any other irrational number. It can solve equations!

I recall the day I received a telegram from Gosper explaining how to construct a glider gun. I gave the telegram at once to friend Bob Wainwright, who had a computer program for exploring Life forms. He put Gosper's glider gun on the screen, and to our amazement it began shooting off gliders. If you care to learn more about Life, you'll find three chapters on it in my book *Wheels, Life, and Other Mathematical Amusements.*

My first column on Life made Conway an instant celebrity. The game was written up in *Time.* All over the world mathematicians with computers were writing Life programs. I heard about one mathematician who worked for a large corporation. He had a button concealed under his desk. If he was exploring Life, and someone from management entered the room, he would press the button and the machine would go back to working on some problem related to the company! Other cellular automaton games have been invented, but none with the amazing elegance and richness of Life.

On one of his visits Conway gave me his fiendish dissection of a cube into a small number of polycubes and told me it was difficult to solve. He was right. I couldn't solve it. Months later I gave a model to Gosper, who solved it quickly, though not by hand. He simply gave the problem to his computer!

One of Conway's other great discoveries was a completely new way to define "number." It, too, was the topic of a *Scientific American* column. Not only was Conway able to generate all familiar numbers, but his method also produced numbers not previously recognized. Donald Knuth, Stanford's famous computer scientist, was so intrigued by Conway's method of creating numbers that he named them surreal numbers and wrote an entire novel, *Surreal Numbers*, about two archaeology students who dig up some stone tablets on which God tells how to generate such numbers.

I had the great pleasure of introducing Conway to Benoit Mandelbrot, the "father of fractals." Mandelbrot then lived not far away in another Westchester town. He came to my house, where he and Conway discussed matters far over my head. Conway had been making new discoveries about Penrose tiling, and Mandelbrot was interested because Penrose tiling patterns are fractals. You can keep enlarging or diminishing them, always to obtain similar patterns.

My column on Penrose tiles made the cover of *Scientific American*. The cover was actually drawn by Conway. On one of his visits he asked for a ruler, compass, and protractor, and in an hour or so drew the tiling pattern that was later colored by a *Scientific American* staff artist.

Penrose tiles, I should explain, are two tiles, called darts and kites, that tile the plane *only* in a nonperiodic way, or what today is called *aperiodic*. To Penrose's vast surprise, it turned out that three-dimensional forms of his tiles would tile space only aperiodically! Not only that, but such shapes could actually be fabricated in laboratories. They became known as quasicrystals. Hundred of papers have since been published about them. They are a marvelous example of how a mathematical discovery, made with no inkling of its application to reality, may turn out to have been anticipated by Mother Nature!

Another column that produced a flood of mail was my April Fool hoax. (It is reprinted in my book *Time Travel and Other Mathematical Bewilderments*.) In it I introduced what I claimed were stupendous breakthroughs in science and math. They included a map that required five colors, a computer proof that Pawn to King's Rook four is a certain chess win for white, the finding of a sketch by Leonardo da Vinci proving he had invented the flush toilet, a psychic motor that rotated when a hand was held near it, and so on. I received hundreds of letters showing how to color my map with four colors. Many readers, including a few scientists, thanked me for alerting them to such important discoveries but chided me for being totally mistaken about one of them. A mathematician phoned to tell me I should be expelled from the American Mathematical Society for not revealing in the May issue that the column was a joke.

My column on Newcomb's decision paradox, which I won't explain here, produced so many letters from readers claiming to have solved the paradox that I took them all to Robert Nozick, a Harvard philosopher, who had been the first to discuss Newcomb's paradox. I persuaded him to write a guest column about reader responses to my hoax, including an amusing letter from Isaac Asimov. Nozick concluded that the paradox is still unresolved. If you care to learn about this marvelous paradox, see chapters 13 and 14 of *Knotted Doughnuts and Other Mathematical Entertainments*.

Another famous mathematician I met through my column was Donald Knuth, now retired from Stanford. His series of books, *The Art of Computer Program-*

*ming*, have made him the world's best-known computer scientist. Because Knuth likes to include in those books as much recreational material as he can cram in, he once visited me at my home in North Carolina. At that time my library and files were in a condominium that I rented solely to house them. The apartment had a kitchen and bathroom. Knuth stayed there for a week going through my files, leaving a stack of papers he wanted copied and sent to him. He cooked his own meals and on Sunday walked to a Lutheran church not far away.

Knuth, a devout Lutheran, has discussed his faith in a book titled *Things a Computer Scientist Rarely Talks About*. His other book on religion is *3:16*. The title refers, of course, to the most-quoted of all New Testament verses, from the Gospel of John: "For God so loved the world that he gave his only begotten son. . . ." Knuth had the interesting notion of using this verse as a way to sample the entire Bible! He simply checked 3:16 in each biblical book, then wrote a commentary on the chapter, weaving it around the verse. For each chapter he asked a calligrapher to letter the verse in some elegant way. The results were so beautiful that not only do they appear in the book, but framed copies toured the United States as an exhibit!

The floor of Knuth's house, I am told, is a map of the area where he lives. If a visitor wants to know how to get to some desired spot, the furniture is shifted so the relevant portion of the map can be consulted.

Still another famous mathematician who became a friend was the graph theorist Frank Harary. I devoted a *Scientific American* column to his generalization of tic-tac-toe. The object of TTT is to get three of a symbol, say, X or O, in a straight line on a 3 × 3 matrix. In other words, to be the first to form a straight tromino. Why not make the objective, Harary asked, be any polyomino on a field of any size? This at once opened up a vast, hitherto unexplored field of two-person "achievement or avoidance games," as Harary liked to call such games. The field contains many still-challenging unsolved problems.

It was soon after I started my *Scientific American* column that I became a member of a semisecret stag club called the Trap Door Spiders. It had been started by a group of New York City science-fiction writers who wanted a chance to meet regularly in spots where they were away from their wives. The name was based on the habit of a spider species that constructs a hole in the ground and covers it with a trapdoor to shut off the entrance. Charter members included such SF writers as Isaac Asimov, Lester del Rey, Sprague de Camp, Lin Carter, and a miscellaneous assortment of editors, scientists, even an Anglican priest. New members are voted in after a member dies or moves too far from the city to attend meetings. I can't recall who suggested I become a member, but so I was, and it was a great honor.

The Spiders met monthly for dinner either at a member's apartment or at a city restaurant. They take turns as hosts. After dinner a guest, chosen by the host, is put on the "hot seat" for an hour or so to answer questions posed by the members. The guests are anyone the host thinks would be of interest to members.

One of the great delights of my membership was getting to know Isaac, who seldom missed a dinner. His series of mystery stories about a club he named the Black Widow Spiders are based on the Trap Door Spiders. Isaac first sold his Black Widow mysteries to *Ellery Queen's Mystery Magazine*, then later gathered them into books.

Each story follows the same whimsical pattern. The dinner guest is always a person with an unsolved mystery. The mystery could be a murder mystery or any sort of problem. Club members bring their expertise to bear on the problem, coming closer and closer to solving it until finally Henry, the club's waiter, pulls together all the clues and solves the mystery!

On one of my turns as host I invited Stefan Kanfer, then book editor of *Time*, as my guest, and soon we made him a member. I have more to say about Steve in chapter 18.

Asimov mentions the club briefly in his autobiography. He cites me as having persuaded him to write annotated editions of his favorite books, starting with Byron's *Don Juan*—books modeled after my *Annotated Alice*, which began a rash of similar volumes by other writers.

I recall a visit to Isaac's apartment on Sixth Avenue overlooking Central Park. I noticed that he worked in a room with no windows. This was by design. If the room had windows, he said, he would be tempted to leave his typewriter to look out the window, and that would seriously interrupt his thoughts while composing.

Isaac was one of the world's best writers of both science fiction and books about science. Asimov was a devout atheist. I once asked him if he had any desire after death to live again. He assured me he had not. I think he lied.

# 16

## PSEUDOSCIENCE

Extraordinary claims require
extraordinary evidence.

—*Carl Sagan*

My reputation as a debunker of bad science began
with an article in the *Antioch Review* titled "The Hermit Scientist." It was about crank scientists who work
in isolation from genuine scientists. A high school
friend, John Eliot, was then living in New York City
as a literary agent. He came across my article, realized he had known me in Tulsa, and came to see me.
He thought my piece could be expanded to a book,
and persuaded me to give it a try. I went to work and
soon had enough of a manuscript for John to take to
a publisher.

To my surprise John quickly sold the manuscript
to Putnam's, who issued it as a hardcover under the
title *In the Name of Science*. It had chapters on Dianetics (before L. Ron Hubbard enlarged it to the religion
of Scientology), Wilhelm Reich's orgonomy, flying
saucers, Hollow Earthers, chiropractors, homeopaths,
and other examples of science moonshine.

The book had poor sales and was soon remaindered. To my delight, Dover picked it up and published a soft-cover edition with the new title *Fads and Fallacies in the Name of Science*. It quickly became one of Dover's top sellers.

Young Colin Wilson, newly renowned in England for his book *The Outsider*, came to see me a few years later. I assumed it was because he agreed with my attacks on the cranks I wrote about. No, it was because I had introduced him to so many fascinating fringe scientists. He talked nonstop about his opinions while I listened in silence. At one point I said, "Colin, the trouble with you is that you're not God." I expected him to be insulted, but he nodded and said I had made a perceptive remark. A week or so later I got a letter thanking me and Charlotte for our hospitality and adding that our discussions were the most stimulating he had had in years. I doubt if he considered a thing I said. Wilson went on to write forgettable novels, and many books praising pseudoscience and the occult, including a hagiography of Reich and a volume extolling the psi powers of Uri Geller.

The person who contributed most to the sales of Dover's *Fads and Fallacies* was Long John Nebel, an all-night radio talk-show host. Long John's specialty was interviewing cranks. Every night, for months, he had someone on the show defending some sort of scientific baloney—persons who claimed to have been abducted by aliens in UFOs, followers of Hubbard and Reich, eye exercisers, psychics, and so on. Each night my book was viciously attacked. As a result, sales soared!

*Fads and Fallacies* led to my becoming a friend of Hayward Cirker, founder and head of Dover. I later wrote introductions to many Dover books, especially fantasies by Gilbert Chesterton, Lewis Carroll, and L. Frank Baum. A pal from my Chicago days, Everett Bleiler, author of numerous books about science fiction and fantasy, was in want of a job. I recommended him to Hayward, who hired him as an editor, a position he held for many years.

I also introduced a magic friend, Karl Fulves, to Hayward, which resulted in a spate of Dover books by Karl on conjuring. Karl also published his own books on magic, including my two booklets exposing the methods of Uri Geller, the Israeli magician turned psychic. My Geller books pretend to be the diary of a fake psychic named Uriah Fuller. I'm surprised I was never sued by Uri. Perhaps he realized that in court he would have to demonstrate his ability to bend spoons and perform other great psychic feats, and in court there would be magicians present to expose his methods. Uri must have known I wrote the Uriah Fuller books because he once called me a "bedbug."

One evening near midnight, when I was half-asleep and changing Jimmy's diaper, I turned the radio on to listen to Long John. The first words I heard were "Mr. Gardner is a liar." The words came from John Campbell, editor of *Astounding Science Fiction*. He was referring to my having written somewhere that Dianetics had failed to cure his sinusitis.

Dianetics was the brainchild of an eccentric science-fiction writer, L. Ron Hubbard. It claimed to be a new and superior therapy that can be described as an unintended caricature of Freudian psychoanalysis. Freud traced mental ills back to childhood experiences. Hubbard went further back. Mental ills, he maintained, are caused primarily by experiences in the womb! Before an embryo develops ears, it records on its cells everything the mother says and hears! The recordings are called engrams. The aim of therapy is to eliminate engrams by a question-and-answer technique called auditing.

Dianetics was first inflicted on the world by an article in John Campbell's *Astounding Science Fiction* that praised Hubbard's 1950 book *Dianetics*. Years later Hubbard added to the therapy a wild, idiotic mythology involving deities and the reincarnation of our thetans (immortal souls). You can read about all this nonsense in the chapter on Dianetics in *Fads and Fallacies*, and a later chapter on Hubbard in my *The New Age*.

The cult soon captured the minds of many simpleminded Hollywood stars, who donated fortunes to the cult. It flourishes to this day in spite of stiff competition from the latest Hollywood fad, the ancient Jewish mystical tradition called the Kabbalah, currently promoted by that great philosophical scholar, Madonna.

My chapter on Dianetics placed me for a time on Hubbard's "suppressive persons" list. Owing to my criticisms of Dianetics I was sent a letter saying I was now "fair game." This meant that any Scientologist could do anything he or she liked to harass me, the "enemy." In her book *The Scandal of Scientology* (Tower Publications, 1971), an early attack on the cult, Paulette Cooper states she was almost killed by over ten years of attacks, including bomb threats. I turned over my "fair game" letter to the FBI. Luckily nothing bad happened to me.

*Fads and Fallacies* was the first of a raft of books by others that debunk current pseudosciences. I continued to do the same in articles and book reviews. Philosopher Paul Kurtz, aided by psychologist Ray Hyman, sociologist Marcello Truzzi, magician James Randi, and myself founded an organization called the Committee for Scientific Investigation of Claims of the Paranormal (CSICOP). It has since shortened its name to CSI. Truzzi edited its first official periodical, the *Zetetic*.

After a few issues of this magazine, it became apparent that Truzzi and other founding members of CSICOP were on different wavelengths. Truzzi didn't care much for debunking. He wanted the magazine to be a scholarly periodical that did not take firm stands but tried to present alternative points of view. For example, if we published an attack on Velikovsky's mad cosmology, we should allow Velikovsky or one of his followers space to reply. We, however, considered Velikovsky the very model of a crackpot, only a small cut above a Flat Earther, who did not deserve to be treated like a reputable scientist. In brief we believed in taking strong sides against crazy science. Marcello did not.

Marcello, recognizing our disagreements, resigned from CSICOP. He kept the title of his magazine, the *Zetetic*, and we began a new periodical called the *Skeptical Inquirer*, under the skilled hands of Kendrick Frazier, a former editor of *Science News*. The new magazine soon grew into a handsome bimonthly that reports the latest news about pseudoscience. Marcello formed his own rival organization with an advisory board that included several well-known cranks. Marcello, by the way, was the son of a world-famous juggler who was for years with the Barnum & Bailey Circus. "I'm not the world's best juggler," he once said, "but the most entertaining."

Carl Sagan, who wrote two excellent books attacking pseudoscience, was lambasted by his colleagues in astronomy for wasting time on cranks. We members of CSICOP agree with Sagan. We believe it the *duty* of scientists to debunk bad science. A democracy works best when citizens are enlightened voters. Reputable science took a heavy drubbing under George W. Bush, especially in his opposition to stem-cell research, and his urging that creationism be taught in public schools as a viable theory alongside evolution! Only harm can result from an electorate unable to tell good science from bogus. This is particularly true with respect to alternative medicines such as homeopathy, which have no support among "allopaths," as homeopaths like to call reputable doctors.

Back now to John Campbell. John may have been a good SF editor, but his ignorance of science was monumental. For example, in many issues of *Astounding Science Fiction* he plugged a psychic machine invented by someone named Hieronymus. Isaac Asimov, in his autobiography, calls it a device of "surpassing idiocy." Campbell's political views were even worse. Asimov describes them as "to the right of Attila the Hun."

When Dover issued the paperback of *Fads and Fallacies*, the *Village Voice* gave away copies to new subscribers. Letters of protest flooded the *Voice* from irate Villagers, some of them sitting nude in Reich's orgone boxes to build up their orgone potential. Many New Yorkers canceled their subscription to the *Voice* to protest my chapter on Reich. The letters subsided after the *Voice* published a letter in which I revealed that I had sent my chapter on Reich to Reich himself before the book was published. He responded with high praise, making only one minor correction of a date. If you read my chapter on Reich carefully, you'll see that nowhere do I criticize him. The closest I came was saying he was either a crank or the greatest physicist of modern times. Reich had no objection to this statement!

Reich was arrested for shipping his orgone boxes across state lines with literature claiming they could cure cancer. During a trial he was his own foolish lawyer. Convicted, he died in prison, a martyr in the eyes of followers. Many still promote his books, even the long pipes he claimed could shoot orgone energy into storm clouds to produce rain!

Each year an untold number of people die as a result of putting their trust in Christian Science or some other form of medical crap such as homeopathy. A homeopathic medicine is a drug that homeopaths believe has great curative power if diluted to the point at which only a few molecules or none at all are left in the liquid, salve, or powder. A main dictum of homeopathy is that the more dilute a drug, the stronger its effect! There is a joke about a homeopath who forgot one day to take his pills and died of an overdose.

James (The Amazing) Randi, in his lectures on homeopathy, displays a bottle of a homeopathic medicine in which a virulent poison is diluted to a few or no molecules. He then drinks the entire bottle. Randi currently runs a foundation in Florida with a lively website on pseudoscience. It has a laboratory, complete with one-way mirrors, for testing self-styled psy-chics and other mountebanks. No one in the world is more skillful than Randi in devising ways to trap swindlers and self-deluded parapsychologists.

A few years ago Randi was asked to investigate a group at the University of Wisconsin (shame on the university!), where researchers were practicing "facili-tated communication" with autistic children. This involves holding the waist or arm of an autistic child while the child types coherent messages. It quickly became apparent to Randi that the facilitators were unconsciously guiding the child's hands. (Psycholo-gists call it the "Ouija Board Effect.") Randi set up a few clever tests that made all this clear. The result? Randi was booted off the premises. One child, while looking around the room—never at the keyboard—typed, "I don't like this man with the beard. Send him away."

Facilitated communication, I regret to say, still appears to flourish at several colleges, where it is extracting money from gullible parents who rejoice in the delusion that a loved child has a fine normal mind struggling to communicate with them.

Randi has no academic training, but he knows more about science than some scientists I have met. He has become a national treasure, fighting a battle likely to last a long time in a nation where polls show that about half its citizens believe in astrology, angels, and Satan, and that the entire universe was created by God in six literal days, just as it says in Genesis!

There is a large overlap in techniques used by magi-cians and the methods used by fake psychics such as Uri Geller. Scientists untrained in the conjuring art of

deception are the easiest people in the world to fool.
That is why Randi, a magician, is so much more capable of detecting fraud than any scientist, including especially parapsychologists. He knows, as few scientists do, all the best ways to set traps for catching psychic swindlers. If you want to know how Geller bends spoons, don't ask a physicist, even if he won a Nobel Prize. Ask me or Randi.

It's surprising how many of the great pseudosciences rest on absurd correlations totally unknown to science. Take phrenology, for instance. It correlates character with the size of head bumps. It was once so widely believed that it generated a vast literature of scholarly books and journals around the world. Walt Whitman published his head chart at the front of *Leaves of Grass*. In England George Eliot had her head shaved so she could get better readings. It was said by skeptics that anyone who believed in phrenology should go have his head examined.

Consider ancient astrology. It assumes a correlation between a person's character and the positions of the sun, moon, and planets on his or her day of birth. Could any belief be crazier? Yet it flourishes today, with astrology columns in almost every newspaper except the *New York Times*. A former president and first lady were true believers!

Palmistry assumes a correlation between events in a person's life and lines on his or her palms. Chiropractic rests on a relation between the state of parts of the spine and most human ills. Physiognomy assumes a relation of character traits to the shape of the nose, ears, chin, mouth, eyebrows, and so on. Hundreds of children may die today because gullible mothers believe that vaccinations cause autism.

I'm reminded of "The Pump Song," a ditty that was popular in 1926 when it was recorded by a group called the Jumping Jacks, and many years later was revived by Louis Prima and his orchestra. The song is credited to composers Richard Whiting, Buddy Fields, and Sammy Lerner. (I don't know who wrote the lyrics.) Only the first four lines are memorable:

It's hard to tell the depth of a well
    By the length of the handle on the pump.
It's hard to gauge a camel's age
    By the curl of the hair upon his hump.

# 17

## MATH AND MAGIC FRIENDS

RON GRAHAM IS ONE OF THE MANY FAMOUS MATHEMA-
TICIANS I got to know well through my *Scientific American*
columns. We collaborated on a column about minimal
Steiner trees that gave me an Erdös number of 2. If a
mathematician shares a byline on a paper with Paul
Erdös, the great number theorist, he gets an Erdös
number of 1. If he shares a byline with someone who
has an Erdös number of 1, he gets an Erdös number of
2, and so on. The Erdös Graph shows how the number
holders are linked. I also got a 2 with Frank Harary, the
noted graph theorist, for a joint paper on a technique
for solving logic problems with directed graphs.

Ron is now retired, but when he was at Bell Labs,
he kept a room in which were stored copies of all the
hundreds of papers Erdös had either written or shared
credit for. The papers had previously been kept by
Erdös's mother. After she died and Ron took over, he
became known around Bell Labs as Erdös's mother.

I met Erdös only once. It was at a luncheon at Bell
Labs, in New Jersey, to which I had been invited by
Graham. At that time Ron was head of the Labs' combinatorial math department. The first thing Erdös said
to me was a question: "When did you arrive?" I started
to check the time on my wristwatch when Ron nudged

me and said, "He means, when were you born?" Erdös liked to talk in a vocabulary all his own. Mathematicians have called it Erdish.

Also at the luncheon was a boy prodigy of ten or twelve who came from Brooklyn. Erdös first asked the lad if he spoke French. To my surprise he nodded. For a while he and Erdös discussed in French a problem involving point set theory, which I failed utterly to understand. But the boy did. I have often wondered what became of him.

Ron, by the way, is a top juggler and gymnast. Before he got his doctorate, he and a friend traveled with a circus as trampoline performers. One afternoon, when I met him for lunch at Bell Labs, Ron greeted me outside a building by walking down long steps on his hands. He also rides a unicycle and can do one-arm handstands. If you see his name on a technical paper with Fan Chung, Fan is his wife.

Although when I wrote the *Scientific American* column I worked at home, I attended monthly luncheons for the staff at the Harvard Club, where publisher Gerard Piel was a member. There would always be an honored guest. Piet Hein was one such guest. John Conway was another. (The club didn't object to his coming in sandals.) Dennis Sciama, the eminent British astrophysicist, was still another guest. He had been a staunch advocate of the steady-state theory of the universe, championed by Fred Hoyle and others. At the luncheon he revealed that he had changed his mind and now accepted the rival theory of the Big Bang, Hoyle's derisive term for the theory. "I was long weaseling out of evidence for the steady state," Sciama told us, "until I ran out of weasels."

I cannot now recall if I met Persi Diaconis through a mutual interest in mathematics or in magic. It was probably magic. Persi, a professor of mathematics and statistics at Stanford University, is a top card magician. He was one of the first to master what is called the Zarrow shuffle, invented by Herb Zarrow. It looks exactly like an ordinary riffle shuffle but leaves all the cards unaltered. It's an extremely difficult shuffle to do properly, one I have never been able to master.

The story of how he got into Harvard is worth my telling once more. When I first met him, Persi was an undergraduate majoring in math at Manhattan's City College. He was both a knowledgeable and a skilled card magician. During summers he worked on ships as a poker hustler. When he told me of his desire to get a Harvard doctorate, I recalled that Fred Mosteller, head of Harvard's statistics department, was an ardent magic buff. Indeed, his picture had recently appeared on the cover of the *Linking Ring*, a journal devoted to conjuring.

I wrote to Fred telling him about Persi's desire to enter Harvard. I asked if he could get my young friend into Harvard, adding that he did the cleanest second and bottom deals of any magician I knew. It worked. Mosteller wrote back to ask if Persi was willing to work for a higher degree in statistics. "Of course," Persi said.

A trip to Harvard for an interview, most of which I suspect was spent with a deck of cards, clinched the deal. Soon Persi was at Harvard, where he obtained a doctorate. Persi is the author of a raft of technical papers and coauthor with Ronald Graham of a book titled *Magical Mathematics: The Mathematical Ideas That*

*Animate Great Magic Tricks.* He is best known for his proof that seven riffle shuffles are necessary and sufficient to produce a randomly ordered deck. Like James Randi, a friend of us both, Persi has long been a foe of parapsychologists who deceive themselves into thinking there is solid statistical evidence for the reality of ESP, PK, and precognition.

More recently Persi was in the news for having shown that when someone flips a coin many times, there is a very slight probability, greater than one-half, that the coin will fall with the same side up as at the flip's start. The reason is that occasionally a flipped coin *looks* like it is spinning when actually it only wobbles. Persi is among the few magicians who can toss coins that wobble.

You'll find Persi's portrait and a brief account of his work in a big beautiful book titled *Mathematicians*, edited with photographs by Mariana Cook (Princeton, 2009). "Mathematicians are exceptional," Ms. Cook opens her preface. "They are not like everyone else. They look like the rest of us, but they are not the same." My book *Last Recreations* is dedicated to Persi as follows:

> *To Persi Diaconis for his remarkable*
> *contributions to mathematics and conjuring,*
> *for his unswerving opposition to psychic*
> *nonsense, and for a friendship going back to our*
> *Manhattan years.*

It was with great reluctance that I decided I had reached a point where it was impossible for me to keep writing *Scientific American* columns and at the same time finish books I wanted to write. Moreover,

I was starting to age and felt it was time for someone younger than I to take over the column. Thereafter, Douglas Hofstadter, newly famous for his book *Gödel, Escher, Bach*, wrote a column called Metamagical Themas for nearly three years. *Scientific American*'s Brian Hayes then developed a column called Computer Recreations, which Canadian computer scientist A. K. Dewdney took over. Finally, Ian Stewart, a British mathematician, contributed a column called Mathematical Recreations.

Most of the good friends I made on the *Scientific American* staff—editor Dennis Flanagan, for example—have since passed away. An exception is Joseph Wisnovsky. In 2009, as an editor at Hill and Wang, Joe bought the manuscript of my book *When You Were a Tadpole and I Was a Fish*. The title is the first line of a popular poem called "Evolution," by a New York City journalist named Langdon Smith. Smith is an extreme example of what critic Burton Stevenson called a "one-poem poet."

There are other one-poem poets, but in every case except Smith they have written scores, sometimes hundreds, of other poems that today are totally unremembered. I am thinking of such poems as "Casey at the Bat," "The Night Before Christmas," "The House by the Side of the Road," "Out Where the West Begins," "The Lost Chord," "The Old Oaken Bucket," and so on. Smith is the only one-poem poet I have yet to encounter who is not known to have published *any* other poem!

The entry on Smith in *Who's Who*, which he may have written himself, does mention that in addition to "Evolution" he wrote a poem called "Bessie McCall of Suicide Hall." Over the years I've tried desperately to track this down, with zero success. My chapter on Smith updates an essay I wrote long ago for the *Antioch Review*. Almost as remarkable as the fact that there is no other known published poem by Smith is the fact that my essay is the only known article about him! He deserves better than that. Finding his poem about poor Bessie would be a literary event that at least one person—namely, me—would find a great achievement.

The most famous fan of my column, to my vast surprise, was Salvador Dalí. On two occasions, when he was in New York City, he invited me to lunch. On the first visit he had with him a copy of my *Ambidextrous Universe*. Also present at the luncheon was a strange but attractive young woman who called herself Ultra Violet. At the time she was trying to organize a rock band, and she needed a name for the group. I suggested "The Infra Reds" but she thought that too obvious. She had been getting lots of publicity out of her crazy insistence that she came from Venus. It turned out she was looking for someone to write her biography. Dalí thought I might be interested!

I asked if she had a publisher for her life story. She said no. I had not the slightest desire to write her biography. However, the lunch ended pleasantly, and I have often wondered what finally became of Ultra. She faded quickly from the scene after posing nude for an erotic tabloid.

The second time Dalí took me to lunch he had with him his beautiful wife, Gala. She appears in many of Dalí's paintings, notably in *Corpus Hypercubus*, where she is looking up at Jesus hanging on a cross that is a hypercube unfolded into eight unit cubes.

Dalí had an intense interest in mathematics. His *Last Supper* swarms with proportions in golden ratio. He produced paintings of landscapes that become faces when rotated ninety degrees, and anamorphic art that comes to life when viewed in a cylindrical or conical mirror. He did paintings in duplicate that become three-dimensional when viewed in mirrors that blend the two pictures.

After our first lunch I walked with Dalí down Fifth Avenue to a Marlboro bookstore he wanted to visit. Of course every few minutes pedestrians, recognizing Dalí's famous mustache, would stop him for an autograph, which he would hastily scribble on a piece of paper handed to him with a pen.

At our second lunch I gave Dalí a small porcelain figure that was a duck in one position and a rabbit in another. Soon after our meeting, the *New York Times* ran a full-page ad for a plastic ashtray Dalí had designed for one of the airlines to give passengers. Around the ashtray were three swans. When you inverted the ashtray, the swans turned into the heads of elephants. I like to think my little rabbiduck had inspired the ashtray.

After our second lunch I had the pleasure of kissing Gala. I can't recall now how this came about. My kiss was a peck on one cheek. Not knowing the French custom, Gala had to explain that now I must kiss her *other* cheek. I made the mistake later of asking her what had happened to Ultra Violet. Gala snapped back, eyes blazing, and said, "You'll have to ask Salvador."

A good friend of my Manhattan days was Gershon Legman, best known as the compiler of a huge book of blue limericks. The book was first published in France, where it was never copyrighted. Convinced it would do well in the United States, I called it to the attention of Nat Wartels, then head of Crown and one of my publishers. Immediately interested, Nat got in touch with Legman, then living with his wife in Valbonne, France, and bought rights to the book. Gershon followed with the two volumes of his *Rationale of the Dirty Joke*.

I first met Gershon through a mutual interest in origami. Gershon was fascinated by paper folding. He sponsored in Paris what was probably the world's first exhibit of origami creations.

Legman introduced me to Lillian Oppenheimer, who did more than anyone to promote origami both here and in Japan. The Japanese had almost forgotten the art. It was kept alive only by the geisha girls. Lillian visited Japan, where she located the nation's leading origami expert, then living in abject poverty. A wealthy widow, Lillian arranged for him a permanent income. She appeared on Japanese television to promote paper folding. I should add that by a previous marriage Lillian had three sons, all of whom became top mathematicians: Martin Kruskal, at Princeton University, Joseph Kruskal, at Bell Labs, and William Kruskal, University of Chicago.

Lillian organized what was probably the first exhibit in the United States of origami figures. It was held at New York City's Cooper Union. On opening day I had the immense pleasure of meeting a daughter of the great Spanish philosopher, poet, and novelist Miguel de Unamuno. As my readers will know, Unamuno is one of my heroes. It is not widely known that he was a skillful paper folder who made significant contributions to the art. I had in the exhibit a paper bird that balanced horizontally if you put a finger tip under its beak. The secret: a penny concealed inside each wing tip.

Gershon had early in life worked for the Kinsey Institute. For a while he edited *Neurotica*, a magazine devoted to all kinds of erotica. Through its pages he promoted his expression "Make love, not war." Gershon died in 1999, leaving the manuscript of a massive autobiography yet to find a publisher.

One of my good friends of later years was Albert Parry, a Russian opponent of Communism who managed to escape being shot and was able to find his way to the United States. He authored many books, starting with a history of tattooing, including a big book on modern terrorism and a wonderful history of American bohemianism titled *Garrets and Pretenders*.

I first met Albert when he was doing graduate work at the University of Chicago. I was working then on an article about Thornton Wilder, and Parry had earlier written a piece on Wilder. I finally located him, and he allowed me to sit down, read his article, and take notes. Years later, after we met again and became friends, he visited me often and was largely responsible for convincing me that Stalin was an evil tyrant, and that Marx had nothing of value to contribute to modern economics. By coincidence his two sons had the same first names as my two sons—James and Thomas.

Parry and Vladimir Nabokov were friends. Nabokov's novel *Pnin* is based on Parry. One day Albert showed me a set of Nabokov's Russian books that Nabokov had given to Albert, each inscribed, some with little drawings of butterflies. Collecting butterflies was one of Nabokov's two hobbies. The other was chess. One of his novels is about a chess grand master, and Nabokov invented and published numerous elegant chess problems.

I am unable to resist adding that Nabokov's novel *Look at the Harlequins!*, about a man who can't distinguish left from right, was heavily influenced by my book *The Ambidextrous Universe.*\* In that volume I quoted two lines from Nabokov's fine poem that opens his novel *Pale Fire*. As a joke, instead of crediting the lines to Nabokov, I credited them to John Shade, who in *Pale Fire* is the poem's supposed author. In his novel *Ada*, Nabokov returned the joke. On page 542 he quotes the same lines, and adds "as quoted by an invented philosopher ('Martin Gardiner')." I don't know if misspelling my name was accidental or part of Nabokov's joke.

Before I met Charlotte, my social life in New York City was confined entirely to friends in magic. Bruce Elliott, who edited a conjuring magazine, the *Phoenix*, would entertain magic friends every Friday night in his apartment on the Upper West Side. It was through Bruce that I met dozens of magicians, most of whom became friends. I met "The Amazing" Randi, about whom I have much to say elsewhere. I met and became friends with the great Dai Vernon, "The Professor," as he was called, who contributed more to modern magic than anyone. Dai spent his elderly years living at Hol-

---

\* See "The Ambidextrous Universe of Nabokov's *Look at the Harlequins!*," by D. Barton Johnson, in *Critical Essays on Vladimir Nabokov*, ed. Phyllis Roth (G. K. Hall, 1984); and "Ambivalence: Symmetry, Asymmetry, and the Physics of Time Reversal in Nabokov's *Ada*," by N. Katherine Hayles, in her book *The Cosmic Web: Scientific Field Models and Literary Strategies in the Twentieth Century* (Cornell University Press, 1984).

lywood's Magic Castle, where he was a permanent entertainer.

I met Señor Wences, the famous ventriloquist. I met Clayton Rawson, author of several mystery novels about Merlini, a magician detective. Though he was not a magician, I must not forget Tony Ravielli, a friend of Bruce and an artist best known for his work on *Sports Illustrated* covers. Tony illustrated several of my books.

I made friends with Paul Curry, whose "Out of This World" is one of the greatest card tricks ever invented. I got to know Roy Benson, whose stage act, which I saw several times at the Roxy, was a thing of beauty. There were Francis Carlyle, Frank Garcia, Stanley Jaks, Dr. Jacob Daley, Walter Gibson, Oscar Weigle, Bill Simon, Vosburgh Lyons, Howie Schwartzman, and other magicians both professional and amateur, far too many to mention.

I can't recall now how or when I first met Bob Orben and his wife, Jean. Bob began his career as a magic-shop demonstrator and author of a small book of jokes suitable for magicians. It was soon followed by larger joke books, and a periodical of topical humor that went to the nation's politicians. Bob became a comedy writer for Red Skelton, Jack Paar, Dick Gregory, and other entertainers. For a time he was a speech writer for President Ford.

Bob told me one day that his files on jokes were in cabinets so heavy that they buckled the floor. Eventually he became so familiar with formulas for jokes that he seldom had to consult his files. We have kept in touch by correspondence over the years, and by occasional visits.

I met Bill Gresham. Bill was the author of *Nightmare Alley*, the best novel ever written about carnival life. It became a movie starring Tyrone Power as a carny who slides downhill to become a geek, one who bites heads off chickens and eats glass. Bill once said to me that one day he realized that his geek was a symbol for all the persons who bitterly hate their jobs but have no other way to make a living.

Bill had been married to Joy Davidman, both once active members of the Communist Party. Joy was drama and poetry editor of the party's *New Masses* magazine. She finally became disenchanted with Communism and wrote a series of articles for the *New York Post* titled "Girl Communist." As a result of reading books by C. S. Lewis, Joy became a convert to the Anglican Church. She began corresponding with Lewis, then an elderly bachelor in England, best known for his Narnia fantasies for children, and for his many books of Christian polemics.

One day Joy, who had become increasingly bitter about her marriage, said to Bill she was planning to divorce him and to take her two children to London where she planned to marry Lewis!

Bill thought that was the funniest thing he ever heard. Then suddenly, to his vast astonishment, Joy did exactly what she said she would do. She divorced Bill, took her sons to England, and sent word to Bill that she and Lewis would soon be married!

After the marriage Lewis published a book titled *Surprised by Joy*. The title of course had a double meaning—the joy that results from faith, and the woman who had suddenly entered a dull life that Lewis once described by saying, "I love monotony."

Now for the punch line. Bill once said to me that the book Lewis wrote was wrongly titled. It should have been *Overwhelmed by Joy*. As all Lewis fans know, the joy was short-lived. Joy developed cancer and soon died. Lewis wrote another book, titled *A Grief Observed*, in which he grapples with the mystery of why God permits so much suffering.

After Joy's death, Bill visited Lewis to arrange for the care of his two young sons. He took with him a copy of my *Annotated Alice*, inscribed to Lewis as a gift. I asked Bill to ask Lewis if he had ever read one of Baum's Oz books. The answer was no. Baum was a great admirer of Carroll's two *Alice* books. He called one of his non-Oz fantasies *A New Wonderland* before the title was changed to *The Magical Monarch of Mo*. It is no coincidence, I believe, that the first word in the first *Alice* book is "Alice," and the first word in the first Oz book is "Dorothy."

Bill was a great admirer of L. Frank Baum. For an issue of the *Baum Bugle* he wrote a moving essay telling of a time when he was deeply depressed over his divorce from Joy. He phoned a friend who allowed him to spend the night in a daughter's vacant bedroom. In the room Bill found a copy of Baum's *The Scarecrow of Oz*, a book he had loved as a child. He spent the night reading the book again. It got him through the night.

When I first met Bill, he paid me a great compliment as we shook hands. "I'm told," he said, "you're a real carny." I knew what he meant. One of my regrets is that as a youth I was never "with it," perhaps as a magician in the Ten-in-One tent.

# 18

## CHARLOTTE

Should the wide world roll away
Leaving black terror
Limitless night,
Nor God, nor man, nor place to stand
Would be to me essential
If thou and thy white arms were there
And the fall to doom a long way.

—*Stephen Crane*

LIKE MOST MEN I HAD MY SHARE OF GIRLFRIENDS before I fell in love—with Charlotte. About these friends I will have nothing to say except that they were wonderful, and I half-loved them all.

Throughout my life I have thought of myself as not being attractive to women. I am a self-centered intellectual, indifferent to all religions, not tall and handsome. I'm five feet eight, and skinny. I have never weighed more than 130 pounds. In one way I resemble one of my heroes, Sherlock Holmes, who, Watson tells us, "loathed every form of society with his whole Bohemian soul." Groucho Marx is credited with the remark that he would never join a club willing to have him as a member. Something close to this view clouded my relationships with the fair sex. I always dimly felt there must be something wrong with any woman who took me seriously.

This, of course, was not entirely true, and there were elements of cruelty if I ended a relationship. I consider this my most besetting sin. I will say no more about my love life on the grounds that I can't imagine any reader having the slightest interest in the details.

Charlotte was an exception. I met her on a blind date arranged by my good friend Bill Simon. Bill was one of many pals in the strange world of conjuring. By profession he worked in his father's New Jersey plant that made brake blocks for cars. Bill was a skilled performer of card magic, well known to magicians for several books on advanced card work, and a hardcover book titled *Mathematical Magic*.

At the time, Bill was dating a girl who had a divorced Bronx friend named Charlotte Greenwald. He arranged for the four of us to have dinner at a restaurant in Manhattan. I remember vividly my first impression of a beautiful young woman with a great smile, lovely green eyes, and a long feather that stood straight up on one side of a hat. When we danced, Charlotte followed without the slightest impulse to lead. And she had a wonderful smell.

Smells are, of course, impossible to describe, but I knew Charlotte's smell was not from any kind of perfume. It came from her hair and body, and clung to her clothes. If someone had presented me with a dozen dresses worn by a dozen women including Charlotte, I could have easily picked out her dress by its smell. I recall now a passage in a book by Somerset Maugham. He wrote about once asking one of H. G. Wells's mistresses what it was about Wells that most appealed to her. He expected her to say his intellect or his humor. She surprised him by saying it was Wells's delicious smell!

Well, enough about olfactory matters. I fell in love almost at once, and for the first and last time. As for marriage, the big hurdle was that I was then very poor. I had saved a small sum from sales to *Esquire*, but *Esquire*, as I explained earlier, no longer cared for my style of writing. The new editor was interested mainly in fiction by famous writers like Hemingway. Charlotte also had small savings from a recent job with J. K. Lasser, a big accounting firm. I was struggling to make a living as a freelance writer, but with limited success. Adding to future bleakness, I was developing cataracts, inherited from both parents, but for me the clouding of lenses came at a much earlier age. On one of our dates Charlotte noticed I had a large hole in the seat of my trousers. Salvation arrived when I became contributing editor to *Humpty Dumpty*, as I have told in a previous chapter.

With what seemed to be a fixed income from *Humpty* we decided to take the plunge. Neither of us wanted a fancy wedding for which we could not pay, but I had a magic friend, George Starke, who was a city judge. He performed the ceremony in his office, at no charge, with Bill serving as best man. Even our Wassermann tests, required by the city, were free. Another magic buddy, Dr. Vosburgh Lyons, a New York City psychiatrist with strong anti-Freudian views, did the tests in his office. A joker, Voss frightened Charlotte by going into another room, then emerging, crouched over, his hair combed over his forehead, chuckling, and brandishing a huge hypodermic needle.

After our marriage we lived for a short time in a third floor walk-up on Fourteenth Street; then we moved to a larger apartment at 17 Charles in the Village. It was in an old brownstone that long ago was replaced by a taller apartment building. Above us lived famous pianist and composer Norman Dello Joio. His cockroaches found their way into one of our closets and had to be professionally exterminated. Below us lived Helen Lawrenson, a writer best known for her memoirs, *Stranger at the Party*, and for an *Esquire* article titled "Latins Are Lousy Lovers." Her daughter, whose name I don't recall, surprised her parents by marrying Abbie Hoffman, a political radical, author of *Steal This Book* and *Revolution for the Hell of It*. Arrested some forty times, mainly on drug charges, he killed himself in 1988.

Our first son, Jimmy, was born while we lived in the Village. In those days the Village was almost free of street crime. There were evenings when I stayed at home as babysitter while Charlotte had no fear of walking alone at night to a nearby movie theater.

Anxious to escape from New York City, and Charlotte pregnant with Tom, our second child, we sold the Village apartment and moved to Dobbs Ferry, a suburb on the Hudson River in New York's county of Westchester. Thomas Owen was soon born. The Dobbs Ferry house, at 26 Bellair Drive, was too small for our enlarging family, and my growing need of more space for books and files. After selling the house we moved a few blocks south on the same street. The move crossed a town line and our address changed to one more appropriate for a mathematician—10 Euclid Avenue—in the town of Hastings-on-Hudson.

It was there that I became friends with a neighbor, Stefan Kanfer, then book editor of *Time*. I reviewed his first book, *A Journal of the Plague Years*. Steve's theme was that the Red menace of the Joe McCarthy era was enormously exaggerated. It had almost no influence on American politics except to frighten the public. A small band of Hollywood Communists did their best to sneak propaganda into films, but with minimal success. McCarthy turned out to be little more than a buffoon. H. L. Mencken, by the way, sounded a similar note in a *Liberty* magazine article titled "The Red Bugaboo." It deserves reprinting.

Steve soon left *Time* to become a successful freelance writer and drama critic for the weekly *New Leader*, a liberal periodical that had once been the loudest voice warning of the evils of Stalinism. Kanfer hit the jackpot with his biography of Groucho, soon followed by *Ball of Fire*, a life of Lucille Ball, and *Somebody*, a life of Marion Brando.

A friend of Steve's, Frank Scioscia, opened the town's used bookstore on the banks of the Hudson. He needed a name for the store. I suggested "riverrun," the first word of *Finnegans Wake*, and that became the store's name. Now run by Frank's son-in-law, in my opinion riverrun is one of the finest rare book stores in New York State.

I think Joyce's *Ulysses* is a great modern novel, but as for *Finnegans Wake*, I consider it a monstrous curiosity of little value. Admirers of the book have wasted incredible amounts of time exploring the novel's thou-

sands of puns and other forms of wordplay. Max Eastman recalls a visit with Joyce at which Joyce proudly
told him he had concealed the names of scores of
rivers in *Finnegans Wake*. In no way, wrote Eastman,
am I going to waste energy trying to locate all those
rivers. It's true that here and there the book breaks
into music, but on the whole his wife described it well
when she called it a mess of chop suey.

My readers have been amazed at how often I found
an appropriate line or two from *Finnegans Wake* to
quote at the top of an essay. They imagine I'm a careful student of the book. I am not. My secret is that I
find such quotes by consulting a mammoth glossary
of *Finnegans Wake* and checking words, or fabricated
words, that in some way relate to an article's topic.

From Hastings-on-Hudson we moved to a house
in Hendersonville, North Carolina. Having by then
abandoned my *Scientific American* column, I no longer
needed to be a short train ride from New York City.
We had friends who had earlier settled in Hendersonville, a sleepy little suburb of Asheville. They had
high praise for the town, so we made a trip there to
check it out and at once fell in love with the village. It
had been the last home of Carl Sandburg, and Asheville was the hometown of Thomas Wolfe, and the
last home of O. Henry. Both Wolfe and O. Henry are
buried there, and the statue of the angel, in Wolfe's
novel *Look Homeward, Angel*, is on display in an Asheville cemetery.

Our first Hendersonville house was far in the country. It had a spectacular view of the mountains through
a picture window, and plenty of room for my files and
books. Its downside was that the nearest supermarket

was a half hour's drive along a narrow twisted road. We sold the house and moved to a new development in the heart of Hendersonville. It had a small lake with two black swans. Late one night, one of the swans crossed a road. I failed to see it, and ran over it. Of course I paid for a replacement, another black swan. At the bottom of the lake lived a large snapping turtle. Residents were never able to locate and dispose of it. I told them I had asked Charlotte if she would go barefoot and locate the turtle by wading around the lake, but she refused.

Once again our house proved too small for my ever-expanding files and library. We moved to our last home, a large Tudor-style stucco, close to the center of town.

While we lived in Hendersonville, our younger son, Tom, lived not far away in Greenville, South Carolina. where he struggled to earn a living as an artist. He had graduated from the Rhode Island School of Design, and his ambition was to become a successful painter. Unfortunately, all his pictures are abstract to the degree that it is difficult to tell their merit. An occasional sale to a local gallery kept hopes alive, and he added to those sales by taking on jobs such as house painting.

From Greenville Tom later moved to Asheville, closer to Hendersonville, and is living there today. He is unusually bright, reads lots of science fiction and other books, and writes great letters. I partially support him, and take him as a dependent on my income tax.

Our older son, Jim, graduated at Kenyon College with a double major in psychology and anthropology, then went on to obtain a Ph.D. in education at the University of Michigan. His doctorate was on the use of computers to help the intellectually disabled. He married Amy, a Kenyon classmate, and as I write, Jim is a professor of special education in the educational psychology department at the University of Oklahoma. Amy is a school psychologist in the public schools. He and Amy have three talented children: Martin, William, and Kate. Martin plays the trumpet, William is an actor, and Kate is an actress and singer.

For a while Jim was interested in magic and actually performed at children's parties. Like most professors, he has liberal political and economic views, as does his wife, and like almost all psychologists, he doubts the reality of paranormal forces as such ESP, PK, and precognition, and such fake therapies as facilitated communication for autistic children. In high school he starred in a production of *The Pirates of Penzance*. His knowledge of technology is immense, and that goes for my grandson Martin as well. I'm online with an Apple Cube Jim handed down, and although I refuse to use email, Google has become an indispensable research tool. When my Apple crashes, Jim is on hand to fix it.

You may wonder why my two sons were not brought up as either Jewish Charlotte's background) or Methodist (mine). Before we were married, I said I would be happy to join a synagogue, but Charlotte refused. She said she would be happy to join a Methodist church, but I refused. In my final chapter I explain why Charlotte and I called ourselves "philosophical theists," with no interest in any organized faith, but who nevertheless believed in God and hoped for an afterlife.

When we first looked over Hendersonville as a place to live, Charlotte described it as a "pretty how town," quoting from the first line of a poem by E. E. Cummings:

anyone lived in a pretty how town
(with up so floating many bells down)

Indeed, Hendersonville has a church with bells floating in its belfry. My friend Ron Edge, a retired physicist at the University of South Carolina, made frequent visits to Hendersonville to ring its bells. We always got together to exchange information about recreational physics, a topic of interest to us both. Ron would write up details about new science tricks and toys to buy in the periodical *Physics Teacher*.

Another physicist friend interested in the same field is Donald E. Simanek, of Lock Haven, Pennsylvania. He runs a great website devoted to the topic. Don is especially interested in ingenious perpetual motion machines that look on paper as if they might work, but when models are built, their wheels refuse to turn. (See the chapter on perpetual motion in my book *The New Age*.)

In her elderly years Charlotte had two hobbies that brought her much pleasure. The first was collecting porcelain birds, a collection she eventually sold to a local antiques dealer. The other was collecting old iron doorstops. Many of them are exquisite examples of miniature sculpture, such as a set of characters from Lewis Carroll's two *Alice* books. Charlotte acquired a magnificent collection, which she finally auctioned off through a firm run by Jeanne Bertoia, author of the first handbook on collecting doorstops.

Charlotte's collecting was done mainly by our visiting antique fairs within driving distances. The trips were our little vacations. Charlotte would look for doorstops, while I looked for copies of Hugo Gernsback's monthly periodical *Science and Invention*, and *John Martin's Book*, a magazine for children. I finally obtained a complete run of both periodicals. Later I sold my *Science and Invention* magazines to a Gernsback collector, and I gave my *John Martin* magazines to the University of North Carolina, at Asheville. You'll find my tribute to Gernsback and his magazine, the great delight of my youth, in my book *From the Wandering Jew to William F. Buckley, Jr.* My tribute to *John Martin's Book* is in the same volume.

After Charlotte died in 2000, and I was unable to pass a test for renewal of my driver's license, I realized that at eighty-six it was time to check into an assisted living facility. I chose Norman because Jim is nearby. He takes care of my shopping needs and visits once a week. I occupy a single-room apartment at Windsor Gardens, in Norman, where I am given five pills every morning after breakfast. My major health problem is type 2 diabetes, which is under control by medication. My blood pressure is low, my cholesterol is so-so, and my vision is perfect. I had cataract operations when I was in my forties and have never had a change in a lens prescription since. As I type, I was ninety-five on October 21, 2009.

My room is large enough for one file cabinet and a small number of books. The rest of my library is in storage along with other file cabinets. I donated my math books to Stanford University at the request of Donald Knuth, the eminent computer scientist. My files on pseudoscience have gone to Prometheus Books, in Amherst, New York, a house that has published a dozen or so of my books attacking bogus science. They also keep in stock my comic religious novel, *The Flight of Peter Fromm*. It's about a Pentecostal young man who, while attending the University of Chicago's Divinity School, gradually loses his faith, but retains belief in God and an afterlife.

My other novel, *Visitors from Oz*, seems to be for children but actually is for adults, especially adults who are Oz fans. It tells of visits by Dorothy and her pals, the Tin Man and the Scarecrow, to some strange towns in the purple Gillikin region of Oz. I assume that Glinda has moved Oz to a parallel world. Dorothy and her two friends, by way of a curious topological structure called a Klein bottle, are able to travel to New York City to publicize a new musical about Oz. The *New York Times* called my book "a poor thing of a novel," but the London *Times Literary Supplement* gave it a long and good review. You can find similar reviews on Google.

At ninety-five I still have enough wits to keep writing. I'm far behind friend Asimov's some three hundred books, but I've found time to come close to a hundred if you count booklets under a hundred pages. The count is still higher if it includes books for children, and books for magicians. Like Asimov, I enjoy writing and seldom suffer from writer's block. My most important book is *The Whys of a Philosophical Scrivener* because it is a confessional of all my beliefs. My last chapter here will summarize those prejudices. My second-best book, *The Night Is Large*, is a collection of essays. The title comes from one of my favorite quotations. It's from Lord Dunsany's play *The Laughter of the Gods*: "Man is a small thing, and the night is very large and full of wonders."

Dunsany is little read today, but I have relished his novels and short fiction ever since I discovered his *Last Book of Wonder* in a Tulsa library. My favorite Dunsany novel is *The Blessing of Pan*. It's about a rural English village that, under Pan's influence, abandons Anglicanism to return to paganism.

Two of my other literary heroes are H. G. Wells and G. K. Chesterton. I have written elsewhere that if you can understand how I can admire both men, one a devout atheist, the other a devout Catholic, you can begin to understand my brand of theism. Each man has for me a positive and a negative side. Wells was blind to the possibility there could be a God and an afterlife, but he had great respect for and understanding of science. I am fond of Wells mainly because of his visions of utopia—a conviction that humanity is now capable of creating a world free of wars and injustices. I am fond of Chesterton because of his sense of humor, his writing style, his fantasies, above all his constant amazement and gratitude that we and a universe exist. More about this in my final chapter.

# 19

## BOB AND BETTY

I first met Bob Murray at Camp Mishawaka, the summer camp my brother and I attended. Decades later my two sons and three grandchildren would also go there. Jim and his wife, Amy, later became Mishawaka counselors.

Like friend John Shaw, Bob was raised a Catholic. Unlike Shaw he later abandoned the faith. The three of us formed a kind of triumvirate that lasted throughout our lives.

Betty Mitchell, a few years younger than I, lived across the street from the house where I grew up. I don't remember how she and Bob met. It's possible I introduced them. At any rate, soon after they met, Betty informed Bob and me that she had discovered an amusing display in the window of a Tulsa department store. A female mannequin had a right hand in the air with its middle finger extended upward. Bob and I drove at once to the store to check. Sure enough, Betty's description was accurate!

One hot summer afternoon, after Bob and Betty began dating, the pair were driving somewhere with Bob at the wheel. He was wearing a pair of short-short shorts. He told me later that he was so aroused by Betty sitting close to him that the tip of his penis crawled into view beyond the edge of his shorts.

Bob's marriage was a Catholic ceremony to please his mother. Betty was not Catholic but had no objection to a Catholic wedding. I recall Bob telling me that before the ceremony, it was obligatory that he attend confession. All he could think of to say was that he had been having "dirty thoughts."

For a while Bob and Betty lived in an apartment near the University of Chicago while I lived a few blocks away. Bob was in Chicago to edit a periodical called *Traffic World*. The publisher was a Catholic friend of Shaw who owned a Chesterton collection that rivaled Shaw's. From *Traffic World* Bob moved to an editorial post with *Advertising Age*, also edited in Chicago.

A few years later he moved to New Jersey to edit in Manhattan a Time-Life series that dealt with home decorating. I was then living in New York City so I continued to see Bob and Betty, and their two daughters, Susan and Betsy, though much less often than when we all lived in Chicago.

I come now to my reason for writing this chapter. It is to record for posterity a few of Bob's harmless practical jokes. I swear that all of them are true and not like so many funny practical jokes that are fabricated events that never took place.

For a brief period, before his marriage, Bob attended classes at Columbia University. One evening, coming from a bar and slightly under the influence, Bob crossed the campus past the statue of Rodin's *Thinker* that stands in front of Columbia's philosophy building.

This statue has always amused me because it seems as if the university is telling the world that *here* is where we have all our deepest thinkers.

On an impulse, Bob climbed to the top of the statue and sat on its head, where he assumed the statue's pose of a nude man deep in thought. As students walked by and saw someone perched on the statue's head, of course they stopped to stare. Bob then announced that they could ask him any question and he would give them a wise answer. Alas, no recording was made of the questions and answers. After a while Bob climbed down and stumbled away. He told me that the next day he went back to the statue and was unable to figure out how he had managed to get to the top. At the time the climb seemed easy and effortless. I later wrote a short story based on the incident, but no magazine would take it, and I long ago threw it away. The editors probably thought I had invented the event.

One day when Bob was visiting an amusement park alone, he took a ride on a concession that had two seats opposite a long bar that rotated so anyone in either seat was periodically upside down as the bar turned. Bob had a newspaper with him. While the bar spun, he opened the paper and calmly pretended to read. A crowd noticed this and gathered to watch. After the ride ended, the man running the ride told Bob if he cared to repeat the newspaper bit, he could have the next ride free. Bob declined and wandered away.

One afternoon when Charlotte and I were visiting my parents in Tulsa, Bob happened also to be in town.

We ran into him on Main Street. He asked where we were heading, and we named Tulsa's largest department store. We parted, walking in opposite directions. After a couple of blocks we entered the store and took an elevator to the third floor. As we stepped off the elevator, Bob greeted us with a grin as he silently entered the elevator for a ride down. We had no memory of telling Bob what we intended to buy, but we must have told him. He had then rushed to the store by an alternate route, took the elevator to the floor where the store carried whatever we intended to buy, then waited by the elevator for us to arrive! It caught us by total surprise. The incident is typical of Bob's uncanny ability to invent good jokes on the spur of the moment.

I had a Tulsa friend named Bing Soph who also was a friend of Bob. Bing and his family moved to Houston. One evening Bing gave a party attended by former Tulsans. While the party was in full swing, there was a knock on the door and a strange-looking man with long whiskers and dark glasses asked if he could join the party. He was welcomed inside, and an hour or so went by before the stranger pulled off his beard and removed his glasses to reveal that he was Bob Murray, whom everyone knew. Not a person there had suspected it was Bob.

Bob's interest in practical jokes went back to his high school days. He and a friend had stationery printed with a meaningless pattern at the top beside the name of a club. Bob had drawn the pattern with his eyes shut. The two used the stationery for a series of jokes of which I will describe only one. A letter was sent to a manufacturer of paper clips. It said they had counted the clips in a box they had purchased, and there were only ninety-eight clips, although the box said it contained a hundred. Moreover, when they first opened the box, the letter continued, a strange odor had emanated from the box. Believe it or not, Bob showed me the letter they got from the paper clip company. It said that when a box was filled with clips, sometimes there were fewer than a hundred, sometimes a trifle more. As for the peculiar odor, they had no idea what had caused it.

One of Bob's funniest jokes involved a Chicago burlesque theater. Bob and a Tulsa friend, Bob Griffith, put on dark glasses and hung on their necks a sign that said, "Help the Blind." They then entered the theater and took seats in the front row. Murray later told me that when a dancing chorus noticed the two men in the front row, they were so cracked up that they got their steps confused.

Bob's red-haired wife, Betty, never participated in Bob's jokes, but she enjoyed telling about them. And he always had a supply of great jokes, most of them on the blue side, which she told in a way that rivaled the best of stand-up comics. She liked to tell of an occasion when she and Bob were in their bathroom, and Bob was sitting on the side of the empty tub with a martini in one hand. He slipped and fell into the tub, then managed to get back on his feet without spilling a drop of the drink!

My favorite Murray joke was a masterpiece. It involved the visit of a Chicago lady friend to their New Jersey home. She stayed overnight. After she left, the Murrays found a single glove she had left behind. Instead of sending her the glove, Bob had a fiendish idea. He took the glove to several department stores until he found a perfect match. He then mailed the lady a glove, but it was of the *same handedness* as the one she had! Bob told me they never heard from her again.

# 20

## GOD

"God is dead."—Nietzsche
"Nietzsche is dead."—God

—*Anonymous graffiti*

WHEN MANY OF MY FANS DISCOVERED THAT I BELIEVED in God and even hoped for an afterlife, they were shocked and dismayed. They seemed to think that if I doubted Uri Geller could bend spoons with his mind, I must be an atheist! Allow me here a chapter to clarify what I mean by the word *God*.

I do not mean the God of the Bible, especially the God of the Old Testament, or any other book that claims to be divinely inspired. For me God is a "Wholly Other" transcendent intelligence, impossible for us to understand. He or she is somehow responsible for our universe and capable of providing, how I have no inkling, an afterlife.

As a great admirer of Rudolf Carnap—with whom, by the way, I once wrote a book—I regard all metaphysical and theological assertions as in a sense meaningless, including the statements made in the previous paragraph. I certainly don't "know" there is a God or an afterlife. I can only hope there are both, and with that hope travels a dim belief.

Faith, in those familiar words of Saint Paul, is the "substance of things hoped for, the evidence of things not seen." I haven't the foggiest notion of what God is like or how an afterlife is possible. In my *Whys of a Philosophical Scrivener* I give three models of an afterlife and conclude that none of them is true. Is God inside or outside time? Is He one or many? How could I, with my tiny brain, possibly know?

The Christian God is a trinity. The Hindu God is a quaternity. For all we know, there could be a hierarchy of gods. Atheists like to taunt theists by asking, "Who created the creator?" Maybe He always existed. Maybe not. There is a wonderful tale by Lord Dunsany, "The Sorrow of Search," in his collection *Time and the Gods*. It tells of a King who climbs a high mountain in search of gods greater than the "Gods of Old"—Asgool, Trodath, Skun, and Rhoog.

Soon the King comes upon a trinity of great deities. Climbing higher, he encounters other great gods until he finally reaches what he believes is the Ultimate God. While his companion, a Master Prophet, is carving on a rock at the end of their search, he sees in the distant haze the faces of four even greater gods. Perhaps you guessed their identities. Yes, they are Asgool, Trodath, Skun, and Rhoog!

Dunsany's circle of deities is based on what mathematicians call a "nontransitive relation." The four Gods of Old, call them A, are below B, B is below C, and so on until the circle reaches A, the four Gods of Old! Even William James, a theist, on the last page of his *Varieties of Religious Experience*, considers the real possibility of polytheism.

Indeed, why limit one's faith to one God? The best answer is Occam's Razor. Our heart has no need to multiply deities that are not needed. It is difficult enough, in an age of reason and science, to believe in one God. Unless you grew up in a culture such as India, where there is a deeply entrenched polytheism, a single God is sufficient to satisfy the heart's hunger. True, the ancient Greeks and Romans got along fine with their beautiful gods, and so do the people of India, but for those of us in the Western world who feel the need for a meaning behind the universe, and who hope for an afterlife, a single deity is enough.

I agree with Carnap that religious beliefs have almost nothing to do with reason. They are based on emotions of fear, and a hope that we and our loved ones will not vanish forever, devoured by Lewis Carroll's Boojum.

This emotive theory of religion, as it is sometimes called, drives atheists up a wall. In a chapter on Rousseau in his *History of Western Philosophy*, Bertrand Russell writes that he prefers Thomas Aquinas to theists like Rousseau because Aquinas at least gave arguments. Philosophical theists, of the sort I admire, *have* no good arguments. It's as if I maintained that my late wife was beautiful and you thought her ugly. Reasonable debate is hopeless.

I can't believe, for emotional reasons, that this fantastic universe popped into existence all by itself. Think about it. At the moment of the Big Bang, some thirteen billion years ago, you and I were there *in potentia*. The laws of physics were such that after billions of years a lonesome universe would evolve creatures as curious as you and me. (To gain energy we have to push organic material through a hole in our face. Sex is even funnier.) If an atheist like John Dewey ever felt wonder at this existence, or for that matter at the existence of a daisy, I can't recall a single passage in which he expressed it.

Russell was once asked what he would say if he found himself standing next to God's throne. He replied, "I would ask God why he didn't give us evidence." For reasons I don't understand, perhaps I'm incapable of understanding, God wants *uncompelled* faith. Maybe some day I will understand. Or maybe not.

# 21

## MY PHILOSOPHY

I don't want to achieve immortality
through my work. . . . I want to achieve
it by not dying.

*—Woody Allen*

I DECIDED TO END THESE DISHEVELED MEMOIRS WITH A
few words about my basic philosophical opinions.
In the introduction to Gilbert Chesterton's *Heretics*,
he contends that the most important thing to know
about anyone is his or her fundamental beliefs. Here
is the passage:

> But there are some people, nevertheless—and I am one
> of them—who think that the most practical and impor-
> tant thing about a man is still his view of the universe.
> We think that for a landlady considering a lodger, it is
> important to know his income, but still more important
> to know his philosophy. We think that for a general about
> to fight an enemy, it is important to know the enemy's
> numbers, but still more important to know the enemy's
> philosophy. We think the question is not whether the
> theory of the cosmos affects matters, but whether, in the
> long run, anything else affects them.

On economic and political issues I call myself a democratic socialist. Let me hasten to add that I agree with Milton Friedman and his wife, Rose, when they said in their book *Free to Choose* that since the days of FDR America has been a democratic socialist state. In an appendix, the Friedmans list the planks in the platform of socialist of Norman Thomas the last time he ran for president. Every plank, vigorously opposed at the time by conservatives, is now accepted by most Republicans! I also agree with Friedman that all the world's democracies are democratic socialist nations. They are mixed economies, partly free enterprise, partly government controlled. They differ only on the degree of control. Until recently America had the fewest controls, the three Scandinavian countries the most. Only in the United States has socialism become a dirty word.

In chapter 9, out of place chronologically, I recall a symposium in New York City at which Norman Thomas spoke. Thomas is one of my heroes. When our nation jammed thousands of loyal Japanese Americans into concentration camps during the Second World War, Thomas was the only well-known political leader whose voice was raised in protest. Never, like so many gullible liberals, did Thomas fall for the bogus socialism of the Soviets. Most people, Thomas liked to say, don't know the difference between democratic socialism, communism, and rheumatism. Where are the Norman Thomases of today? Obama? Hillary? Of course they can't call themselves democratic socialists.

More memories of the symposium float into my head. Another of its distinguished speakers was Wystan Auden, the British poet. He had little to say, but apparently enjoyed saying it because he laughed constantly at his remarks.

Questions from the audience were written down and handed to democratic socialist Sidney Hook, the moderator. After reading a question intended for Daniel Bell, another democratic socialist, Hook, to the room's amusement, answered the question himself! At this point Bell stood up, turned his seat the other way, and sat with his back to the audience, suggesting his nonpresence as a speaker. I can recall these details vividly although I no longer remember the topic of the debate.

In addition to the great poets of the past, whose poems ring like bells, I am also fond of many poets not among the greatest, but whose poems I believe will outlast, say, the vapid verse of William Carlos Williams. I have edited for Dover two anthologies of popular verse, some great, some not, any one of which I would rather read again than a poem by Ezra Pound.

As for art, my tastes, like my tastes in poetry, are classical. Of most modern art, realistic or abstract, I have a low opinion. Picasso, who straddled the two types of art, produced some splendid paintings when he tried hard, and some abstractionists—Paul Klee, for instance—paint pictures that are pleasing and amusing. But there are many nonobjective painters, whose works sell after their death for millions, whom I consider humbugs devoid of talent. Once for fun I dripped a Pollock parody. I had to take it off a wall because it embarrassed visitors who thought it genuine. Imitation Pollocks are now all over the place, confounding art critics who have no way to distinguish genuine Pollocks from good counterfeits.

The key to success in today's mad art world is to have a gimmick no other artist has, such as Pollock's dribblings, the fat black brush strokes of Franz Kline, the two rectangles of Mark Rothko, or the all-black (or some other color) canvases of Ad Reinhardt. (See my short story "The Great Crumpled Paper Hoax" in *The Jinn from Hyperspace* about the discovery of a new minimal sculpture gimmick.)

Modern sculpture has hit similar lows. I'm thinking of Carl Andre's pile of bricks, of giant replicas of things like a toothbrush, and thousands of other follies that find their way into galleries and parks. The *New York Times* (March 27, 2009) ran a photo of a bronze cat by Swiss sculptor Alberto Giacometti, to be auctioned by Sotheby's on May 5. Alberto cast eight of these miserable cats, each not only devoid of aesthetic merit, but positively ugly. If I found one of his cats on sale in some shop, and didn't know the sculptor's fame, I wouldn't have paid a dollar for it.

Sotheby's expected *Le chat* to sell for $16 million to $24 million. To the great surprise of the auction house, no one wanted to bid for it, and it went unsold.

For a moment I thought this a sign that art collectors were starting to come to their senses, but no. At the same auction was a Mondrian titled *A Composition in Black and White, with Double Lines.* Parallel lines, like railroad tracks, a ruler's width apart, that crossed one another. It sold to an unidentified bidder for $9.2 million. Anybody with a minimum sense of composition, and with a ruler and ballpoint, could have produced this "painting" in fifteen minutes. New York City's art world continues to be as demented as ever in spite of the deep recession then afflicting the nation.

In 1961 Giacometti cast six more than life-sized bronze stick figures titled *L'Homme qui marche* (Man Walking). By stick figure I mean a three-dimensional version of the figures children draw in the margins of textbooks to produce motion when the pages are flipped. Giacometti's skeleton is shown in midstride, his long arms dangling vertically, his head tiny. Other editions of *L'Homme* are in top galleries, a first edition at the Carnegie Museum of Art, in Pittsburgh.

On February 3, 2010, the statue's second edition, cast in 1961, was auctioned at Sotheby's, in London. In eight minutes an anonymous wealthy phone bidder grabbed the statue for sixty-five million pounds, or more than a hundred million in U.S. dollars! It was the largest price ever paid for an auctioned work of art. The "natural equilibrium" of the man's stride, Giacometti once said, symbolizes his "life force." Sotheby's described the Walking Man as "both a humble image of a man, and a potent symbol of humanity." No rhetoric is funnier than the rhetoric of art critics. The statue is pictured on the Swiss hundred-franc banknote.

In ethics I hold to what is called the "emotive theory." I agree with Kant that science can *describe* how people behave, but can't provide an "ought" without the aid of axioms held for emotional reasons. An obvious example of such a posit is "It is better to be alive than dead." I once saw a sign in a Tulsa bar that read, "It is better to be rich and healthy than to be poor and sick."

Now for my religious opinions. I prefer not to call myself a Christian even though I enormously admire almost all (an exception is a belief in hell) of Jesus's teachings)—teachings that deist Thomas Jefferson gathered in his famous "Jefferson Bible." However, I no longer think Jesus was God incarnate, a belief I consider essential for anyone who calls himself or herself Christian.

Somehow, like Peter in my comic novel, I managed to retain faith in a personal God and a hope for an afterlife. I'm what is called a "philosophical theist." It has a distinguished tradition that starts with Plato and includes Kant, Bayle, the deists, William James, Charles Peirce, Ralph Barton Perry, Edgar Brightman, above all, Miguel de Unamuno, the Spanish philosopher, poet, and novelist.

Philosophical theism is based unashamedly on posits of the heart, not the head. It freely admits that atheists have all the best arguments. There *are* no proofs of God or of an afterlife. Indeed, all experience suggests there is no God. If God exists, why would he so carefully conceal himself? All experience suggests that when we die, our body rots and nothing in our brain survives.

After my four years in the navy I returned to the University of Chicago, where the G.I. Bill covered tuition for the most influential philosophy course I ever took. It was a seminar by Rudolf Carnap on the philosophy of science. I was so impressed that later I persuaded Carnap, the next time he gave the course, to have it tape-recorded by his wife. From the recordings she typed out everything Carnap said, along with questions by students and Carnap's answers. Each week she mailed me transcripts, and I shaped them into the only nontechnical work written by Carnap. Basic Books published it under the title *Philosophical*

*Foundations of Physics*. A Dover reprint altered the title to *An Introduction to the Philosophy of Science*. Every idea in the book is Carnap's, every sentence mine. The collaboration was one of the happiest tasks of my life.

Carnap is out of fashion these days among young philosophers, but I believe his reputation will grow again and eventually surpass the crumbling fame of his chief rival, Karl Popper. Popper went out of his way to disagree with Carnap, only to sneak Carnap's opinions back under a different terminology. "The distance between us is an asymmetric function," Carnap once wrote. "From Popper to me the distance is short. From me to Popper the distance is long."

I sometimes call myself a "theological positivist" in the following sense. I agree with Carnap's positivism that religious statements are cognitively meaningless, unsupported by logic or science. I don't *know* there is a God. I don't *know* there is an afterlife. However, a belief in either is not emotionally meaningless. As Carnap once put it, religious beliefs and scientific beliefs are like two separate continents, with no land joining them. One may hope there is a God and another life, but it's a hope resting only on desire.

Shortly before he died, Carl Sagan wrote to tell me he had just reread my *Whys*, and was it fair to say I believed in God only because it made me happier? I responded by saying, in effect, "You've got it!" My faith rests entirely on desire. However, the happiness it brings is not like the momentary glow that follows a second martini. It's a lasting escape from the despair that follows a stabbing realization that you and everyone else are soon to vanish utterly from the universe. It is an effort, perhaps genetic, to relieve the anguish of believing the universe is nothing more than the tale of a blind idiot, full of sound and fury and signifying nothing.

William James once said that his famous essay "The Will to Believe" should have been called "The Right to Believe." In this classic defense of the leap of faith James spells out the conditions that justify such a jump. For example, the decision must be nontrivial, a leap that profoundly affects one's life. You don't, for example, leap to a belief in Santa Claus or Uncle Sam. The jump isn't justified if it is contradicted by science or any other kind of reliable evidence. It isn't justified if it is a belief in, say, UFOs or a monster living on the far side of the moon. The leap must be momentous, a decision that profoundly affects your life.

This emotive approach to faith infuriates atheists like Bertrand Russell. Why? Because there obviously is no way to refute a person's decision to make the leap. There is no way to settle disagreement.

Today the old design arguments for God are enjoying a revival based on the fine-tuning of some twenty physical constants. Let a constant vary by the tiniest fraction and no galaxies could form, no suns, no planets, hence no sentient creatures. Alas, this argument is easily refuted by the possibility there is an infinity of universes, each with random values for its constants, therefore . . .

The logic of this new argument from design is indeed shattered by the concept of a multiverse with varying laws and constants, but at what a price! The argument has strong emotive force. It seems much simpler and more satisfying to believe in one creator God, for which there is no evidence, than to believe in lots of worlds, perhaps an infinity, for which there also is no evidence.

Carl Sagan was more honest. He said it would be wonderful if he could live again, to reunite with parents and other loved ones, and (let me add) to learn about new discoveries of science, such as maybe finding life on other planets. Sagan believed such hopes impossible because of a total lack of evidence.

Along with Unamuno I go a step further. Not only is there no evidence for God or another life, but the evidence strongly suggests the nonexistence of both. The enormity of irrational evil implies the absence of a just God. The brain's deterioration after death suggests that nothing in it survives. The leap of faith clearly is Quixotic, against all odds. One of Unamuno's finest books is his commentary on *Don Quixote*, a book about a deluded man whom Unamuno takes to be a symbol of the person of faith.

In addition to being a philosophical theist, I also belong to a group of thinkers who call themselves "mysterians." We are convinced there are truths as far beyond our brain's capacity now to comprehend as the truths *we* understand are beyond the grasp of a chimpanzee. Although chimps are our nearest relatives (almost all the two DNAs are identical), there is no way to make a chimp understand, say, the square root of 2. We mysterians believe it the height of hubris to suppose our brains will not improve as humanity evolves.

Prominent among today's mysterians are Noam Chomsky, Roger Penrose, and philosophers Colin McGinn, John Searle, and Thomas Nagel. McGinn is the leader of the group and author of several books defending mysterianism, notably *The Problem of Consciousness* (1993). The name "mysterian," by the way, was taken from the name of a rock band. McGinn recently likened the fabrication of self-awareness by that small lump of matter inside our skulls to such incredible events as discovering that a lump of bread was fabricating the counting numbers, or that a system of ethics had emerged from rhubarb. His remark echoes a statement by Thomas Huxley. The appearance of consciousness in a human brain, he said, is as startling as the appearance of a Jinn when Aladdin rubbed his lamp. Kant and Leibniz were other early mysterians. See Google for a good article on the "New Mysterians," and for British journalist John Derbyshire's essay on why he abandoned Christianity to become a secular mysterian.

One deep riddle, far beyond our present understanding, is the mystery of consciousness and free will, two names for essentially the same thing. It is impossible to imagine one without the other. Some mysterians think this mystery is only temporary. Some fine day soon neuroscientists will unravel the riddle. I belong to the more radical wing of mysterianism—those who think this cannot happen for a long, long time, perhaps never.

There are, of course, even darker mysteries. Why does *anything* exist? Why is the universe mathematically patterned to make science possible? Do we fully understand the nature of matter?

Far from it! We know matter is made of molecules, that molecules are made of atoms, and atoms are made of electrons, protons, and neutrons. Protons and neutrons are made of quarks, and quarks and electrons may be made of vibrating loops of superstrings. And what are the strings made of? They seem to be made of pure mathematics. A friend once said that the universe seems to be made of nothing, yet somehow it manages to exist. It is even possible that matter has an infinity of levels. If Kant were alive today, he would regard particle physics as confirming his notion of the *Ding an sich*, the unknowable "thing in itself."

A British writer who had an enormous influence on my thinking is Gilbert Keith Chesterton. Although I am not tempted by his powerful rhetoric in defense of Rome, I read G. K.'s *Orthodoxy* every few years and always with pleasure. I think his *Man Who Was Thursday* is a masterpiece of philosophical fantasy. My fondness for Chesterton's fiction, especially his Father Brown mysteries, is expressed in my book *The Fantastic Fiction of Gilbert Chesterton*. Above all, I love reading anything by G. K. because of his never-ceasing emotions of wonder and gratitude to God, not only for such complicated things as himself, his wife, and the universe, but for such "tremendous trifles" (as he once called them) as rain, sunlight, flowers, trees, colors, stars, even stones that "shine along the road / That are and cannot be," as he has it in his great religious poem "A Second Childhood."

In the atheist periodical *Free Inquiry* (April/May, 2009) Phil Zuckerman has an article titled "Aweism." It is about his preference for calling himself an "aweist" rather than an atheist, agnostic, or secular humanist. His essay is a superb defense of an atheist's right to experience awe—a profound sense of wonder at the great mysteries of existence, a wonder that at times can be close to terror.

Zuckerman's epigraph is a familiar statement by Einstein about how a person without a sense of wonder is as good as dead, like a candle with its flame snuffed out. Is Zuckerman aware that Einstein, in conversation with Max Jammer (see Jammer's *Einstein and Religion*, page 48), stated bluntly that he was neither an atheist nor a pantheist? He was, he said, a firm believer in what he liked to call "the Old One"—a transcendental intelligence, vastly superior to ours, who is responsible for the Big Bang and the slow evolution of at least one universe.

Two words are conspicuously missing from Zuckerman's passionate defense of aweism. The words are "Gilbert Chesterton."

There is, however, one enormous difference between Chestertonian wonder and Zuckerman's awe. Chesterton always combined his wonder with gratitude for the privilege of being alive. The odds against Zuckerman's having been born are astronomical. Had a different pair of sperm and egg united, Zuckerman's mother would have given birth to someone else. If my Aunt Minnie had a mustache, she would have been Uncle Fred. If Zuckerman had ever combined his awe with gratitude, he would have strayed dangerously close to theism. As G. K. liked to say, one of the saddest moments for an atheist is when he feels a deep gratitude for something and has no one to thank.

Consider the glowworm. There is a poem by Lord Dunsany (in his *Fifty Poems*) that goes as follows:

As I walked in a night of July
   On hills where the foxes pass,

The stars were clear in the sky
   And the glow-worms shone in the grass.

And my fancy attuned my ear
   To the voice of the things of the brake;
Their tiny tones I could hear,
   And I heard when a glow-worm spake

And told, in the ear of the elves,
   The scorn of a worm for a star:
"They are glow-worms just as ourselves,
   Only less important by far."

Now there is an obvious sense in which the glow-worm's statement is accurate. A star may be important in making life possible on a planet, but considered in itself it is a dead thing—a lifeless glob of swirling atoms, radiating energy in accord with Einstein's famous equation. Its structure is simple and easily understood. Compared to a star, even to a galaxy, a glowworm is infinitely more complex, a living thing. It will be a long time before neuroscientists fully understand its tiny brain. Soon it will turn into a firefly, a transformation more miraculous than the hocus-pocus of Piet Hein's black earth turning into yellow crocus.

As for God and an afterlife, our head tells us both are illusions. An Old Testament psalm (14:1), Unamuno reminds us, does not say, "The fool hath said in his *head* there is no God." God is a hope only of the heart.

Let John Keats have the final word. In his midtwenties, knowing he was rapidly dying of consumption, never to know if he would be remembered for his poems, Keats in a letter has the following sad words:

Is there another life? Shall I awake and find all this a dream? There must be [an afterlife], we cannot be created for this sort of suffering.

Keats was not listening to his head. His head told him the possibility of another life was near zero. He was listening to his heart.

# AFTERWORD:
# MY MOST ELEGANT FRIEND . . .

WHERE TO BEGIN? I'VE REALLY NO IDEA WHERE—OR exactly when—I first met Martin Gardner. I believe that moment may have occurred in the offices of *Scientific American* magazine almost seven decades ago, but it seems that I have always known him. He became such a fixture in my life, such a dependable part of my world; I was so very accustomed to picking up the telephone to call him, or answering a call from him that would always result in an improvement of my knowledge of the universe. Alas, no more . . .

For twenty-five years, Martin wrote the Mathematical Games column for *Scientific American*, gaining worldwide renown for that task. I can testify to that fact because wherever I went in my international travels, I would be mobbed as soon as my acquaintance with Martin came up for discussion.

He also authored some seventy books. I frequently dropped a mention of my friendship with Martin into conversations with academics, and traveling the world as I have done most of my life, I frequently found that some academics rather doubted that I actually knew this legendary figure in person. I recall that when I delivered a lecture to the systems engineers of IBM many years ago, a talk during which I referred to Martin, following my talk I was besieged by a group from the audience who asked me to settle whether or not Martin was an actual individual, or perhaps an amalgamation of Isaac Asimov, Arthur C. Clarke, and maybe a magician colleague of mine, since his writings so frequently touched on the sort of expertise that only such a trio could summon up. They were appropriately amazed and edified when I assured them that this paragon was actually a single person, a real human being, and quite as accomplished as he appeared to be.

At that IBM appearance, I was required to mention to their systems engineers that their Series 370 business machines, which had been in place beginning in 1970, would not be succeeded by a newer series, a change that had been expected. In fact, those 370s continued into the 1990s. I'd hastened to consult with Martin, asking what he knew about the number 370, to see if I might work it into my presentation. "Aha!" he said—thus also inventing a book title—"370 is one of only four possible numbers—aside from zero and 1—that is the sum of the cubes of its own digits. What's the next highest one?" I had no answer and felt like a fool when he told me. It's quite obvious. "And if you're interested in a Spanish connection," he continued, "turn it upside down." I did, and IBM was very happy with the results. I'm sure Martin could have gone on and on with fascinating facts about any other number we'd cared to choose.

In conversation with a lady I know who worked for *Scientific American* before they lost Martin's column, I learned that they'd conducted a short survey of persons who'd renewed their subscription, asking them the reason for that action. To their surprise, they found that a good percentage of the answers involved the Mathematical Games column, to the exclusion of any other.

Another matter on which I was queried from time to time was whether or not Martin actually had academic degrees in mathematics—which he did not. As he once expressed it to me, after beginning his column in *Scientific American*, he sort of learned it as he went along. And I must say that I believe that was true. He always expressed his delight at something that he had just stumbled upon or that had occurred to his agile mind as he applied it to a problem at hand. Indeed, "delight" was a major characteristic of this man's makeup. That enthusiasm certainly carried over into his books and his *SA* column. He was constantly celebrating discoveries, expanding on them, and looking for new ways to communicate them to the public—and especially to young people. He was never happier than when in the company of kids to whom he would present a brainteaser, followed by the "Aha!" phase in which he would provide an answer—usually totally unexpected—that made everything quite clear.

That lucidity of Martin's work made him a great teacher. His weaving of a story might very well have been inspired by his total admiration for the *Alice* stories by Lewis Carroll. He pored over every sentence that Carroll had constructed and extracted from them every sort of nuance he could, and of course recorded his observations in writing, to the delight of his many, many, fans over the years and around the globe. His spectrum of interest was very broad. His coterie of friends included major professional magicians, mathematicians of every sort, philosophers, a few scoundrels, and a sufficient variety of weirdos to round out his perception of the world.

As an atheist myself, I'll admit that I was somewhat surprised that this man was a deist. When I inquired about this apparent lapse of logic, he calmly informed me that he was well aware the atheists had a much better argument than he did, and that in fact he had no supporting evidence for his acceptance of a deity. He said that his decision simply made him "feel more comfortable." Knowing and loving Martin as I did, I easily accepted that fact and somewhat celebrated it. Anything that improved Martin's life also improved mine.

James Gardner, son of Martin Gardner, has sent me a truly wonderful gift, one that I shall always treasure. Forgive me if I'm a bit weepy at the moment, but this memento of my friend of many decades has affected me more than I thought it would. I spent a lot of time over the long years I knew Martin visiting him and his wife, Charlotte, at their home in Hastings-on-Hudson, and every time I pulled into that driveway I had to wonder whether he'd chosen the house because of its address: 10 Euclid Avenue. I never asked.

Ah, but the gift I've received . . . It's a 19˝ × 23˝ × 11˝ wooden lectern, worn and scratched, well used and bearing the marks of its long service proudly. On that sturdy desk Martin Gardner wrote most of his books, columns, articles, and letters, either by hand or on an ancient mechanical typewriter, about which I could tell another story, at another time . . . Every visitor to my home has been shown that lectern, which still contains two decks of cards that Martin used when inventing his card tricks. They will never be opened by me. I have volunteered to have the desk returned to the Gardner family—where it belongs—upon my demise.

A photo of Martin faces me as I peck out these comments at my desk, a photo to which I bid, "Good morning, Martin!" as I ease into my office chair each morning, and as I turn in at night, I touch the lectern on the way to my slumber.

At our Amaz!ng Meetings that we hold every year, we of the JREF—the James Randi Educational Foundation—don't hold any sort of memorial to Martin Gardner. That would have embarrassed him hugely, I'm quite sure. His son Jim, calling me to announce his father's demise, added that the will he left behind specified that there be no funeral, and that cremation would be preferred. That's my Martin, and I expected no less. No, at all of our JREF conferences we celebrate the existence of this fine gentleman, one of my giants, a huge intellect, a prolific author, and a caring, responsible citizen of the world. If we can manage it, we'll have balloons and dancing girls—which would have titillated Martin, I guarantee you.

Yes, he's gone away, but his wise words and his great love for reason and compassion will remain with us forever. I loved him dearly, but I leave him to the ages . . .

Oh, and in case you're still working on the answers for paragraph four, the numbers are 153, 371, and 407, and 370 turned over is OLÉ . . . See? Yeah, I missed 371, too.

*James Randi*

# INDEX

works (*cont'd*)
Baum biography, 5–6; L.
Frank Baum book introduc-
tions, 6; *Best Remembered Poems*, xv–xvi; book reviews,
xvi; "Bourgeois Idealism in
Soviet Nuclear Physics," 90;
caricatures, 91; *Cherchez la
femme* (cardboard puzzle),
32; G. K. Chesterton's "The
Coloured Lands" introduc-
tion, 6–8; *Children's Digest*
contributions, 133; college
short stories, 59; *Confessions of a Psychic* series, xiv,
152; "Destiny," 23, 24; "Do
Loops Explain Conscious-
ness?", xxvii, 141; Dover anthologies of popular verse,
197; Dover book introduc-
tions, 152; "The Encyclopedia of Impromptu Magic,"
xii; *Esquire* short stories,
125–27, 175; "An Ethral-
drian Gazes at the Earth,"
24–25; *Fads and Fallacies in
the Name of Science*, xiv, 151–
52, 153, 154, 155; *Famous
Poems from Bygone Days*, 91;
*The Fantastic Fiction of Gilbert
Chesterton*, 33, 205; *Favorite
Poetic Parodies*, 11n; *The
Flight of Peter Fromm*, xix,
53, 60, 61, 62, 86, 92, 105–6,
114, 183, 200; foreword to
Roger Penrose's *Emperor's
New Mind*, 140; "Forgot-
ten gods! Alas, the words
convey," 25–26; Martin
Gardner's *The Whys of a Philosophical Scrivener* (review
of), xxi, 82; "The Great
Crumpled Paper Hoax,"
198; Sam Hair's "The Ballad
of Terrible Mike" illustrations, 91; *Here's New Magic*,

74–75; "The Hermit Scien-
tist," 150; "Hexaflexagons,"
135; *Hexaflexagons and Other
Mathematical Diversions:
The First Scientific American
Book of Puzzles and Games*,
22; "H. G. Wells, Premature
Anti-Communist," 132;
high school poetry, 23–24;
*Hobbies* magazine article,
104; Douglas Hofstadter's
*Gödel, Escher, Bach* (review
of), 141; "The Horse on
the Escalator," 125–26;
"I Am a Mysterian," xxi;
*An Introduction to the
Philosophy of Science*, 201;
"The Irrelevance of Conan
Doyle," 22; *The Jinn from
Hyperspace*, 4, 123, 141, 198;
Stefan Kanfer's *A Journal
of the Plague Years* (review
of), 177; *Knotted Doughnuts
and Other Mathematical
Entertainments*, 145; *Last
Recreations*, 163; letter to
*New Republic* of December
13, 1940, 43–44; "Logic
Machines," 135; *Martin
Gardner Presents*, xii; *Math-
ematics, Magic, and Mystery*,
xvi, 134; "Mr. Smith Goes
to Tulsa," 132; *In the Name
of Science*, 150; *Never Make
Fun of a Turtle, My Son*, 131;
*The New Age: Notes of a Fringe
Watcher*, 58, 153, 181; *New
Leader* reviews and articles,
96, 132; *The Night Is Large*,
183–84; "The No-Sided
Professor," xi, 126–27; *The
No-Sided Professor, and Other
Tales of Fantasy, Humor,
Mystery, and Philosophy*, 116,
127; "Occam's Razor," 128;
*Order and Surprise*, 46, 48,